College Algebra

College Algebra

Jonas Hoover

NY Research Press
New York

Published by NY Research Press
118-35 Queens Blvd., Suite 400,
Forest Hills, NY 11375, USA
www.nyresearchpress.com

College Algebra
Jonas Hoover

International Standard Book Number: 978-1-63238-701-1 (Hardback)

Cataloging-in-Publication Data

College algebra / Jonas Hoover.
p. cm.
Includes bibliographical references and index.
ISBN 978-1-63238-701-1
1. Algebra--Textbooks. 2. Algebra--Problems, exercises, etc.
3. Mathematics. I. Hoover, Jonas.
QA154.3 .C65 2019
512--dc23

Printed in China.

Contents

Preface

Algebra is a fundamental area of mathematics that manipulates symbols for solving equations and studying various abstractions. Elementary algebra and abstract algebra are the two primary branches of algebra. Elementary algebra uses variables to denote quantities or general numbers. It is mostly concerned with the solution of linear equations, quadratic equations, exponential and logarithmic equations, radical equations and systems of linear equations. Abstract algebra is the study of algebraic structures such as fields, groups, rings and vector spaces. It is used in many fields of science and mathematics. This book is a compilation of chapters that discuss the most vital concepts in the field of algebra. It attempts to understand the multiple branches that fall under the discipline of algebra. In this book, constant effort has been made to make the understanding of the difficult concepts of algebra as easy and informative as possible, for the readers.

A foreword of all chapters of the book is provided below:

Chapter 1 - Algebra is a fundamental area of mathematics. It includes elementary equation solving and study of abstractions. Some of the important areas of algebra are linear, geometric and multilinear algebra. This is an introductory chapter, which will introduce briefly all the significant branches of algebra; **Chapter 2** - The study of algebraic structures, such as groups, rings, fields, lattices, modules, etc. is under the domain of abstract algebra. The concepts of sets, binary operations, inverse elements and identity elements are fundamental concepts in this field. This chapter has been carefully written to provide an easy understanding of the varied algebraic structures studied under abstract algebra; **Chapter 3** - The field of mathematics that studies algebraic structures is known as universal algebra. The study of variety, clone, structure, post's lattice, etc. is fundamental to this field. This chapter closely examines these algebraic structures to provide an extensive understanding of the subject; **Chapter 4** - Algebraic geometry is a branch of mathematics that studies zeros of multivariate polynomials. Some of the classes of algebraic varieties that are commonly studied are plane algebraic curves, cubic curves and quartic curves. The topics elaborated in this chapter will help in gaining a better perspective about the different concepts that are central to algebraic geometry; **Chapter 5** - A polynomial is an algebraic expression that consists of variables and coefficients. It involves the operations of addition, subtraction, multiplication and exponents. Monomials and binomials are polynomials with one and two terms respectively. This chapter discusses in detail the different kinds of polynomials to develop a comprehensive understanding.

At the end, I would like to thank all the people associated with this book devoting their precious time and providing their valuable contributions to this book. I would also like to express my gratitude to my fellow colleagues who encouraged me throughout the process.

Jonas Hoover

1

Fundamentals of Algebra

Algebra is a fundamental area of mathematics. It includes elementary equation solving and study of abstractions. Some of the important areas of algebra are linear, geometric and multilinear algebra. This is an introductory chapter, which will introduce briefly all the significant branches of algebra.

Algebra

Algebra is a part of mathematics (often called math in the United States and maths or numeracy in the United Kingdom). It uses variables to represent a value that is not yet known. When an equals sign (=) is used, this is called an equation. A very simple equation using a variable is: $2 + 3 = x$ In this example, $x = 5$, or it could also be said, "x equals five". This is called *solving for x*.

Besides equations, there are inequalities (*less than* and *greater than*). A special type of equation is called the function. This is often used in making graphs.

Algebra can be used to solve real problems because the rules of algebra work in real life and numbers can be used to represent the values of real things. Physics, engineering and computer programming are areas that use algebra all the time. It is also useful to know in surveying, construction and business, especially accounting.

People who do algebra need to know the rules of numbers and mathematic operations used on numbers, starting with adding, subtracting, multiplying, and dividing. More advanced operations involve exponents, starting with squares and square roots. Interestingly, many of these rules can also be used on the variables.

Algebra was first used to solve equations and inequalities. Two examples are linear equations (the equation of a line, $y=mx+b$) and quadratic equations, which has variables that are squared (power of two, a number that is multiplied by itself, for example: 2*2, 3*3, x*x). How to factor polynomials is needed for quadratic equations.

Examples

Here is a simple example of an algebra problem:

> Sue has 12 candies, Ann has 24
>
> candies. They decide to share so that they have the same number of candies.

These are the steps you can use to solve the problem:

1. To have the same number of candies, Ann has to give some to Sue. Let x represent the number of candies Ann gives to Sue.
2. Sue's candies, plus x, must be the same as Ann's candies minus x. This is written as: $12 + x = 24 - x$
3. Subtract 12 from both sides of the equation. This gives: $x = 12 - x$. (What happens on one side of the equals sign must happen on the other side too, for the equation to still be true. So in this case when 12 was subtracted from both sides, there was a middle step of $12 + x - 12 = 24 - x - 12$. After a person is comfortable with this, the middle step is not written down.)
4. Add x to both sides of the equation. This gives: $2x = 12$
5. Divide both sides of the equation by 2. This gives $x = 6$. The answer is six. If Ann gives Sue 6 candies, they will have the same number of candies.
6. To check this, put 6 back into the original equation wherever x was: $12 + 6 = 24 - 6$
7. This gives 18=18, which is true. They both now have 18 candies.

With practice, algebra can be used when faced with a problem that is too hard to solve any other way. Problems such as building a freeway, designing a cell phone, or finding the cure for a disease all require algebra.

Writing Algebra

As in most parts of mathematics, adding z to y (or y plus z) is written as $y + z$. Subtracting z from y (or y minus z) is written as $y - z$. Dividing y by z (or y over z: $\frac{y}{z}$) is written as $y \div z$ or y/z. y/z is more commonly used.

In algebra, multiplying y by z (or y times z) can be written in 4 ways: $y \times z$, $y * z$, $y \cdot z$, or just yz. The multiplication symbol "×" is usually not used, because it looks too much like the letter x, which is often used as a variable. Also, when multiplying a larger expression, parentheses can be used: $y\,(z+1)$.

When we multiply a number and a letter in algebra, we write the number in front of the letter: $5 \times y = 5y$. When the number is 1, then the 1 is not written because 1 times any number is that number ($1 \times y = y$) and so it is not needed.

As a side note, you do not have to use the letters x or y in algebra. Variables are just symbols that mean some unknown number or value, so you can use any variable. x and y are the most common, though.

Functions and Graphs

An important part of algebra is the study of functions, since functions often appear in equations that we are trying to solve. A function is like a box you can put a number or numbers into and get a certain number out. When using functions, graphs can be powerful tools in helping us to study the solutions to equations.

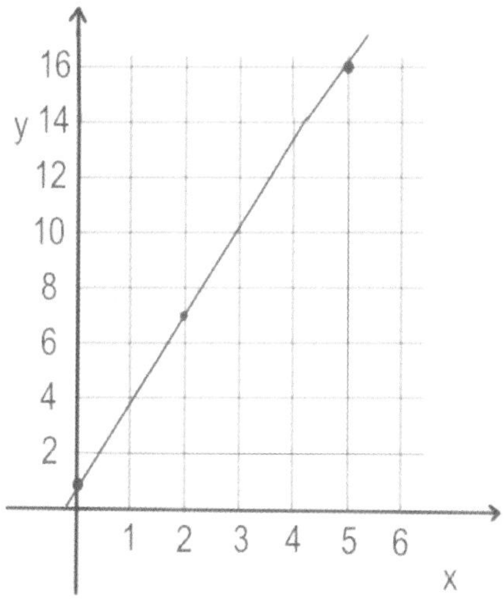

Linear equation for y=3x+1

A graph is a picture that shows all the values of the variables that make the equation or inequality true. Usually this is easy to make when there are only one or two variables. The graph is often a line, and if the line does not bend or go straight up-and-down it can be described by the basic formula $y = mx + b$. The variable *b* is the y-intercept of the graph (where the line crosses the vertical axis) and *m* is the slope or steepness of the line. This formula applies to the coordinates of a graph, where each point on the line is written (x, y).

In some math problems like the equation for a line, there can be more than one variable (*x* and *y* in this case). To find points on the line, one variable is changed. The variable that is changed is called the "independent" variable. Then the math is done to make a number. The number that is made is called the "dependent" variable. Most of the time the independent variable is written as *x* and the dependent variable is written as *y*, for example, in $y = 3x + 1$. This is often put on a graph, using an *x* axis (going left and right) and a *y* axis (going up and down). It can also be written in function form: $f(x) = 3x + 1$. So in this example, we could put in 5 for *x* and get y = 16. Put in 2 for *x* would get y=7. And 0 for *x* would get y=1. So there would be a line going thru the points (5,16), (2,7), and (0,1) as seen in the graph to the right.

If x has a power of 1, it is a straight line. If it is squared or some other power, it will be curved. If it uses an inequality (< or >), then usually part of the graph is shaded, either above or below the line.

Rules of Algebra

In algebra, there are a few rules that can be used for further understanding of equations. These are called the rules of algebra. While these rules may seem senseless or obvious, it is wise to understand that these properties do not hold throughout all branches of mathematics. Therefore, it will be useful to know how these axiomatic rules are declared, before taking them for granted. Before going on to the rules, reflect on two definitions that will be given.

1. Opposite - the opposite of a is $-a$.
2. Reciprocal - the reciprocal of a is $\frac{1}{a}$.

Rules

Commutative Property of Addition

'Commutative' means that a function has the same result if the numbers are swapped around. In other words, the order of the terms in an equation do not matter. When the operator of two terms is an addition, the 'commutative property of addition' is applicable. In algebraic terms, this gives $a+b=b+a$.

Note that this does not apply for subtraction! (i.e. $a-b \neq b-a$)

Commutative Property of Multiplication

When the operator of two terms is an multiplication, the 'commutative property of multiplication' is applicable. In algebraic terms, this gives $a \cdot b = b \cdot a$.

Note that this does not apply for division! (i.e. $\frac{a}{b} \neq \frac{b}{a}$, when $a \neq b$)

Associative Property of Addition

'Associative' refers to the grouping of numbers. The associative property of addition implies that, when adding three or more terms, it doesn't matter how these terms are grouped. Algebraically, this gives $a+(b+c)=(a+b)+c$. Note that this does not hold for subtraction, e.g. $1=0-(0-1) \neq (0-0)-1=-1$.

Associative Property of Multiplication

The associative property of multiplication implies that, when multiplying three or more terms, it doesn't matter how these terms are grouped. Algebraically, this gives $a \cdot (b \cdot c) = (a \cdot b) \cdot c$. Note that this does not hold for division, e.g. $2=1/(1/2) \neq (1/1)/2=1/2$.

Distributive Property

The distributive property states that the multiplication of a number by another term can be distributed. For instance: $a \cdot (b+c) = ab+ac$. (Do *not* confuse this with the associative properties! For instance, $a \cdot (b+c) \neq (a \cdot b)+c$.)

Additive Identity Property

'Identity' refers to the property of a number that it is equal to itself. In other words, there exists an operation of two numbers so that it equals the variable of the sum. The additive identity property states that the sum of any number and 0 is that number: $a+0=a$. This also holds for subtraction: $a-0=a$.

Multiplicative Identity Property

The multiplicative identity property states that the product of any number and 1 is that number: $a \cdot 1 = a$. This also holds for division: $\frac{a}{1}=a$.

Additive Inverse Property

The additive inverse property is somewhat like the opposite of the additive identity property. When an operation is the sum of a number and its opposite, and it equals 0, that operation is a valid algebraic operation. Algebraically, it states the following: $a - a = 0$. Additive inverse of 1 is (-1).

Multiplicative Inverse Property

The multiplicative inverse property entails that when an operation is the product of a number and its reciprocal, and it equals 1, that operation is a valid algebraic operation. Algebraically, it states the following: $\frac{a}{a} = 1$. Multiplicative inverse of 2 is 1/2.

Advanced Algebra

In addition to "elementary algebra", or basic algebra, there are advanced forms of algebra, taught in colleges and universities, such as abstract algebra, linear algebra, and universal algebra. This includes how to use a matrix to solve many linear equations at once. *Abstract algebra* is the study of things that are found in equations, going beyond numbers to the more abstract with groups of numbers.

Many math problems are about physics and engineering. In many of these physics problems time is a variable. Time uses the letter *t*. Using the basic ideas in algebra can help reduce a math problem to its simplest form making it easier to solve difficult problems. Energy is *e*, force is *f*, mass is *m*, acceleration is *a* and speed of light is sometimes *c*. This is used in some famous equations, like f = ma and e=mc^2 (although more complex math beyond algebra was needed to come up with that last equation).

Elementary Algebra

Elementary algebra is the most basic form of algebra taught to students. It is often one of the next areas of mathematics taught to students after arithmetic. While in arithmetic only numbers and operators like +, −, ×, and ÷ occur; in algebra, variables (like *a*, *x*, *y*) are used to stand for numbers. This is useful because:

- It lets people solve problems about "unknown" numbers. This means learning about equations and how to solve them (for example, "find a number *x* where $3x+1=10$").
- It allows the generalization of the rules from arithmetic. While some students understand that $3+4=4+3$, it helps to prove that $a+b=b+a$ for all *a* and *b*. This makes algebra a good step to learning about abstraction (learning general ideas from many examples).
- It helps people understand and create functional relationships (also sometimes called cause and effect). An example of this is "if *x* tickets are sold, then the profit will be $3x-10$ dollars".

These three are the main strands of elementary algebra. Elementary algebra is often used in many other subjects, like science, business, and building. Abstract algebra, a much more advanced topic, is generally taught late in college.

Simple Algebra Problems

If an equation has only one number that is unknown it is sometimes easy to solve. The unknown number is called "x":

$$2x + 4 = 12.$$

To solve a simple equation with one unknown amount add, subtract, multiply, or divide both sides of the equation by the same number in order to put the unknown amount, x, on one side of the equation. Once x is by itself on one side, use arithmetic to determine the amount on the other side of the equation. For example, by subtracting 4 from both sides in the equation above:

$$2x + 4 - 4 = 11 - 4$$

getting:

$$2x = 8$$

Dividing both sides by 2:

$$\frac{2x}{2} = \frac{8}{2}$$

getting:

$$x = 4.$$

Linear Algebra

Linear algebra is a branch of mathematics. It came from mathematicians trying to solve systems of linear equations. Vectors and matrices are used to solve these systems. The main objects of study currently are vector spaces and linear mappings between vector spaces. Linear algebra is useful in other branches of mathematics (e.g. differential equations and analytic geometry). It can also be applied to the real world in areas such as engineering, physics and economics.

Linear algebra describes ways to solve and manipulate (rearrange) systems of linear equations.

For example, consider the following equations:

$$\begin{aligned} x + y &= 0 \\ x - 2y &= 3 \end{aligned}$$

These two equations form a system of linear equations. It is linear because none of the variables are raised to a power. The graph of a linear equation in two variables is a straight line. The solution to this system is:

$$\begin{aligned} x &= 1 \\ y &= -1 \end{aligned}$$

since it makes all of the original equations valid, that is, the value on left side of the equals sign is exactly the same as the value on the right side for both equations.

Linear algebra uses a system of notation for describing system behavior, called a matrix. For the previous example, the coefficients of the equations can be stored in a coefficient matrix.

Multilinear Algebra

In mathematics, multilinear algebra extends the methods of linear algebra. Just as linear algebra is built on the concept of a vector and develops the theory of vector spaces, multilinear algebra builds on the concepts of p-vectors and multivectors with Grassmann algebra.

In a vector space of dimension n, one usually considers only the vectors. According to Hermann Grassmann and others, this presumption misses the complexity of considering the structures of pairs, triples, and general multivectors. Since there are several combinatorial possibilities, the space of multivectors turns out to have 2^n dimensions. The abstract formulation of the determinant is the most immediate application. Multilinear algebra also has applications in mechanical study of material response to stress and strain with various moduli of elasticity. This practical reference led to the use of the word tensor to describe the elements of the multilinear space. The extra structure in a multilinear space has led it to play an important role in various studies in higher mathematics. Though Grassmann started the subject in 1844 with his *Ausdehnungslehre*, and re-published in 1862, his work was slow to find acceptance as ordinary linear algebra provided sufficient challenges to comprehension.

The topic of multilinear algebra is applied in some studies of multivariate calculus and manifolds where the Jacobian matrix comes into play. The infinitesimal differentials of single variable calculus become differential forms in multivariate calculus, and their manipulation is done with exterior algebra.

After Grassmann, developments in multilinear algebra were made in 1872 by Victor Schlegel when he published the first part of his *System der Raumlehre*, and by Elwin Bruno Christoffel. A major advance in multilinear algebra came in the work of Gregorio Ricci-Curbastro and Tullio Levi-Civita. It was the *absolute differential calculus* form of multilinear algebra that Marcel Grossmann and Michele Besso introduced to Albert Einstein. The publication in 1915 by Einstein of a general relativity explanation for the precession of the perihelion of Mercury, established multilinear algebra and tensors as physically important mathematics.

Use in Algebraic Topology

Around the middle of the 20th century the study of tensors was reformulated more abstractly. The Bourbaki group's treatise *Multilinear Algebra* was especially influential — in fact the term *multilinear algebra* was probably coined there.

One reason at the time was a new area of application, homological algebra. The development of algebraic topology during the 1940s gave additional incentive for the development of a purely algebraic treatment of the tensor product. The computation of the homology groups of the product of two spaces involves the tensor product; but only in the simplest cases, such as a torus, is it directly calculated in that fashion. The topological phenomena were subtle enough to need better foundational concepts; technically speaking, the Tor functors had to be defined.

The material to organise was quite extensive, including also ideas going back to Hermann Grassmann, the ideas from the theory of differential forms that had led to de Rham cohomology, as well as more elementary ideas such as the wedge product that generalises the cross product.

The resulting rather severe write-up of the topic (by Bourbaki) entirely rejected one approach in vector calculus (the quaternion route, that is, in the general case, the relation with Lie groups). They instead applied a novel approach using category theory, with the Lie group approach viewed as a separate matter. Since this leads to a much cleaner treatment, there was probably no going back in purely mathematical terms. (Strictly, the universal property approach was invoked; this is somewhat more general than category theory, and the relationship between the two as alternate ways was also being clarified, at the same time.)

Indeed, what was done is almost precisely to explain that *tensor spaces* are the constructions required to reduce multilinear problems to linear problems. This purely algebraic attack conveys no geometric intuition.

Its benefit is that by re-expressing problems in terms of multilinear algebra, there is a clear and well-defined 'best solution': the constraints the solution exerts are exactly those you need in practice. In general there is no need to invoke any *ad hoc* construction, geometric idea, or recourse to co-ordinate systems. In the category-theoretic jargon, everything is entirely *natural*.

Conclusion on the Abstract Approach

In principle the abstract approach can recover everything done via the traditional approach. In practice this may not seem so simple. On the other hand, the notion of *natural* is consistent with the *general covariance* principle of general relativity. The latter deals with tensor fields (tensors varying from point to point on a manifold), but covariance asserts that the language of tensors is essential to the proper formulation of general relativity.

Some decades later the rather abstract view coming from category theory was tied up with the approach that had been developed in the 1930s by Hermann Weyl (by working through general relativity via abstract tensor analysis, and additionally in his book *The Classical Groups*). In a way this took the theory full circle, connecting once more the content of old and new viewpoints.

Topics in Multilinear Algebra

The subject matter of multilinear algebra has evolved less than the presentation down the years. Here are further pages centrally relevant to it:

- bilinear operator
- component-free treatment of tensors

- Cramer's rule
- dual space
- Einstein notation
- exterior algebra
- exterior derivative
- inner product
- Kronecker delta
- Levi-Civita symbol
- metric tensor
- mixed tensor
- multilinear map
- multilinear form
- symmetric algebra, symmetric power
- symmetric tensor
- tensor
- tensor algebra, free algebra
- tensor contraction

Applications

Some of the ways in which multilinear algebra concepts are applied:

- classical treatment of tensors
- dyadic tensor
- bra–ket notation
- geometric algebra
- Clifford algebra
- pseudoscalar
- pseudovector
- spinor
- outer product
- hypercomplex number
- multilinear subspace learning

Geometric Algebra

The geometric algebra (GA) of a vector space is an algebra over a field, noted for its multiplication operation called the geometric product on a space of elements called multivectors, which is a superset of both the scalars F and the vector space V. Mathematically, a geometric algebra may be defined as the Clifford algebra of a vector space with a quadratic form. Clifford's contribution was to define a new product, the geometric product, that united the Grassmann and Hamilton algebras into a single structure. Adding the dual of the Grassmann exterior product (the "meet") allows the use of the Grassmann–Cayley algebra and a conformal version of the latter together with a conformal Clifford algebra yields a conformal geometric algebra (CGA) providing a framework for classical geometries. In practice, these and several derived operations allow a correspondence of elements, subspaces and operations of the algebra with geometric interpretations.

The scalars and vectors have their usual interpretation, and make up distinct subspaces of a GA. Bivectors provide a more natural representation of pseudovector quantities in vector algebra such as oriented area, oriented angle of rotation, toarque, angular momentum, electromagnetic field and the Poynting vector. A trivector can represent an oriented volume, and so on. An element called a blade may be used to represent a subspace of V and orthogonal projections onto that subspace. Rotations and reflections are represented as elements. Unlike vector algebra, a GA naturally accommodates any number of dimensions and any quadratic form such as in relativity.

Specific examples of geometric algebras applied in physics include the spacetime algebra (or the less common alternative formulation, the algebra of physical space) and the conformal geometric algebra. Geometric calculus, an extension of GA that incorporates differentiation and integration, can be used to formulate other theories such as complex analysis, differential geometry, e.g. by using the Clifford algebra instead of differential forms. Geometric algebra has been advocated, most notably by David Hestenes and Chris Doran, as the preferred mathematical framework for physics. Proponents claim that it provides compact and intuitive descriptions in many areas including classical and quantum mechanics, electromagnetic theory and relativity. GA has also found use as a computational tool in computer graphics and robotics.

The geometric product was first briefly mentioned by Hermann Grassmann, who was chiefly interested in developing the closely related exterior algebra. In 1878, William Kingdon Clifford greatly expanded on Grassmann's work to form what are now usually called Clifford algebras in his honor (although Clifford himself chose to call them "geometric algebras"). For several decades, geometric algebras went somewhat ignored, greatly eclipsed by the vector calculus then newly developed to describe electromagnetism. The term "geometric algebra" was repopularized in the 1960s by Hestenes, who advocated its importance to relativistic physics.

Definition and Notation

There are a number of different ways to define a geometric or Clifford algebra. Hestenes's original approach was axiomatic, "full of geometric significance" and equivalent to the universal Clifford algebra. Given a finite-dimensional quadratic space V over a field F with a symmetric bilinear form (the *inner product*, e.g. the Euclidean or Lorentzian metric) $g: V \times V \to F$, the geometric algebra for this quadratic space is the geometric algebra $G(V, g)$, together with the exterior algebra $\wedge V$.

Since it is usual in GA, for the remainder of this topic, only the real case, $F = \mathbf{R}$, will be considered. The notation $\mathcal{G}(p,q)$ or $\mathcal{G}_n$ will be used to describe the geometric algebra.

The essential product in the algebra is called the *geometric product*, and the product in the contained exterior algebra is called the *exterior product* (frequently called the *outer product* and less often the *wedge*). It is standard to denote these respectively by juxtaposition (i.e., suppressing any explicit multiplication symbol) and the symbol $\wedge$. The above definition of the geometric algebra is abstract, so we summarize the properties of the geometric product by the following set of axioms. The geometric product of $\mathcal{G}_n$ has the following properties for multivectors A, B, C:

$AB \in \mathcal{G}_n$

$1A = A1 = A$, where 1 is the unit and A any element of the algebra (existence of an identity element)

$A(BC) = (AB)C$, where A, B and C are any elements of the algebra (associativity)

$A(B+C) = AB + AC$ and $(B+C)A = BA + CA$, where A, B and C are any elements of the algebra (distributivity)

$a^2 = g(a,a)1$, where a is any element of the subspace V of the algebra.(In other words, the geometric product aa is linked to the inner product $a \cdot a$.)

The exterior product has the same properties, except that $a \wedge a = 0$ for $a \in V$.

Note that in the final property above, the real number $g(a,a)$ need not be nonnegative if g is not positive-definite. An important property of the geometric product is the existence of elements having a multiplicative inverse. If $a^2 \neq 0$ for some vector a, then a^{-1} exists and is equal to $g(a,a)^{-1}a$. A nonzero element of the algebra does not necessarily have a multiplicative inverse. For example, if u is a vector in V such that $u^2 = 1$, the element $\frac{1}{2}(1+u)$ is both a nontrivial idempotent element and a nonzero zero divisor, and thus has no inverse.

It is usual to identify $1 \in \mathbf{R}$ with $1 \in G(V,g)$, with associated natural embeddings $\mathbf{R} \to G(V,g)$ and $V \to G(V,g)$. In this topic, this identification is assumed. Throughout, the term *vector* refers to an element of V.

The Geometric Product

For vectors a and b, we may write the geometric product of any two vectors a and b as the sum of a symmetric product and an antisymmetric product:

$$ab = \frac{1}{2}(ab+ba) + \frac{1}{2}(ab-ba)$$

Thus we can define the *inner product*of vectors as

$$a \cdot b := g(a,b),$$

so that the symmetric product can be written as

$$\frac{1}{2}(ab+ba) = \frac{1}{2}((a+b)^2 - a^2 - b^2) = a \cdot b$$

Conversely, g is completely determined by the algebra. The antisymmetric part is the exterior product of the two vectors, the product of the contained exterior algebra:

$$a \wedge b := \frac{1}{2}(ab - ba) = -(b \wedge a)$$

Then by simple addition:

$ab = a \cdot b + a \wedge b$ the ungeneralized or vector form of the geometric product.

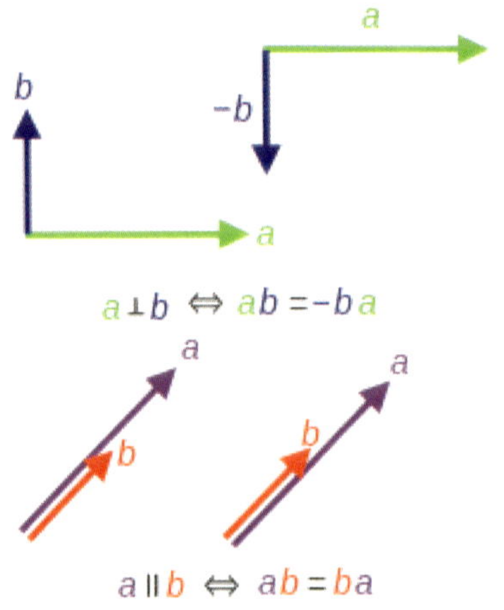

Given two vectors a and b, if the geometric product ab is anticommutative; they are perpendicular (top) because $a \cdot b = 0$, if it is commutative; they are parallel (bottom) because $a \wedge b = 0$.

The inner and exterior products are associated with familiar concepts from standard vector algebra. Geometrically, a and b are parallel if their geometric product is equal to their inner product, whereas a and b are perpendicular if their geometric product is equal to their exterior product. In a geometric algebra for which the square of any nonzero vector is positive, the inner product of two vectors can be identified with the dot product of standard vector algebra. The exterior product of two vectors can be identified with the signed area enclosed by a parallelogram the sides of which are the vectors. The cross product of two vectors in 3 dimensions with positive-definite quadratic form is closely related to their exterior product.

Most instances of geometric algebras of interest have a nondegenerate quadratic form. If the quadratic form is fully degenerate, the inner product of any two vectors is always zero, and the geometric algebra is then simply an exterior algebra. Unless otherwise stated, this article will treat only nondegenerate geometric algebras.

The exterior product is naturally extended as an associative bilinear binary operator between any two elements of the algebra, satisfying the identities

$$1 \wedge a_i = a_i \wedge 1 = a_i$$

$$a_1 \wedge a_2 \wedge \cdots \wedge a_r = \frac{1}{r!} \sum_{\sigma \in \mathfrak{S}_r} \operatorname{sgn}(\sigma) a_{\sigma(1)} a_{\sigma(2)} \cdots a_{\sigma(r)},$$

where the sum is over all permutations of the indices, with $\operatorname{sgn}(\sigma)$ the sign of the permutation, and a_i are vectors (not general elements of the algebra). Since every element of the algebra can be expressed as the sum of products of this form, this defines the exterior product for every pair of elements of the algebra. It follows from the definition that the exterior product forms an alternating algebra.

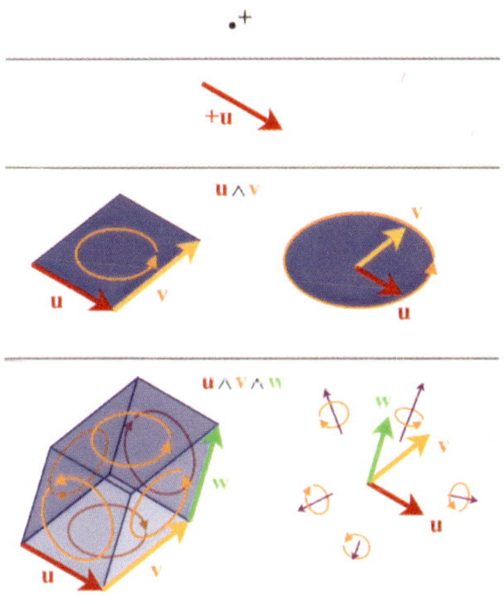

Orientation defined by an ordered set of vectors.

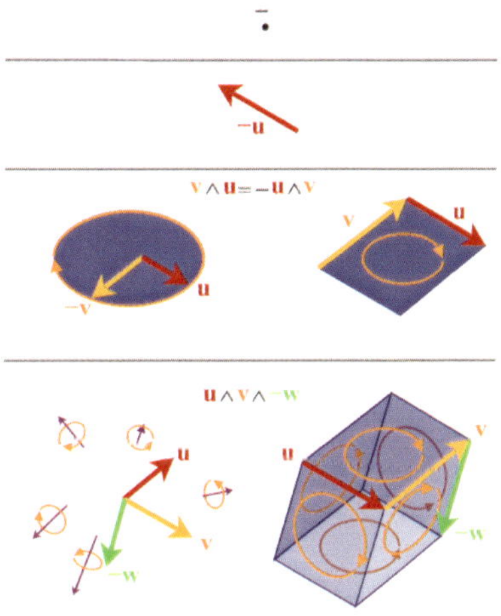

Reversed orientation corresponds to negating the exterior product.

Geometric interpretation of grade- n elements in a real exterior algebra for $n = 0$ (signed point), 1 (directed line segment, or vector), 2 (oriented plane element), 3 (oriented volume). The exterior product of n vectors can be visualized as any n-dimensional shape (e.g. n -parallelotope, n -ellipsoid); with magnitude (hypervolume), and orientation defined by that on its $(n-1)$-dimensional boundary and on which side the interior is.

Blades, Grades and Canonical Basis

A multivector that is the exterior product of r independent vectors ($r \leq n$) is called a *blade*, and the blade is said to be a multivector of grade r. From the axioms, with closure, every multivector of the geometric algebra is a sum of blades.

Consider a set of r independent vectors $\{a_1, \ldots, a_r\}$ spanning an r-dimensional subspace of the vector space. With these, we can define a real symmetric matrix

$$[A]_{ij} = a_i \cdot a_j$$

By the spectral theorem, **A** can be diagonalized to diagonal matrix D by an orthogonal matrix O via

$$\sum_{k,l} [O]_{ik} [A]_{kl} [O^T]_{lj} = \sum_{k,l} [O]_{ik} [O]_{jl} [A]_{kl} = [D]_{ij}$$

Define a new set of vectors $\{e_1, \ldots, e_r\}$, known as orthogonal basis vectors, to be those transformed by the orthogonal matrix:

$$e_i = \sum_j [O]_{ij} a_j$$

Since orthogonal transformations preserve inner products, it follows that $e_i \cdot e_j = [D]_{ij}$ and thus the $\{e_1, \ldots, e_r\}$ are perpendicular. In other words, the geometric product of two distinct vectors $e_i \neq e_j$ is completely specified by their exterior product, or more generally

$$\begin{aligned} e_1 e_2 \ldots e_r &= e_1 \wedge e_2 \wedge \ldots \wedge e_r \\ &= \left(\sum_j [O]_{1j} a_j\right) \wedge \left(\sum_j [O]_{2j} a_j\right) \wedge \ldots \wedge \left(\sum_j [O]_{rj} a_j\right) \\ &= \det[O] a_1 \wedge a_2 \wedge \ldots \wedge a_r \end{aligned}$$

Therefore, every blade of grade r can be written as a geometric product of r vectors. More generally, if a degenerate geometric algebra is allowed, then the orthogonal matrix is replaced by a block matrix that is orthogonal in the nondegenerate block, and the diagonal matrix has zero-valued entries along the degenerate dimensions. If the new vectors of the nondegenerate subspace are normalized according to

$$\hat{e}_i = \frac{1}{\sqrt{|e_i \cdot e_i|}} e_i,$$

then these normalized vectors must square to $+1$ or -1. By Sylvester's law of inertia, the total number of $+1$s and the total number of -1s along the diagonal matrix is invariant. By extension, the total number p of these vectors that square to $+1$ and the total number q that square to -1 is invariant. (If the degenerate case is allowed, then the total number of basis vectors that square to zero is also invariant.) We denote this algebra $\mathcal{G}(p,q)$. For example, $\mathcal{G}(3,0)$ models 3-dimensional Euclidean space, $\mathcal{G}(1,3)$ relativistic spacetime and $\mathcal{G}(4,1)$ a conformal geometric algebra of a 3-dimensional space.

The set of all possible products of n orthogonal basis vectors with indices in increasing order, including 1 as the empty product, forms a basis for the entire geometric algebra (an analogue of the PBW theorem). For example, the following is a basis for the geometric algebra $\mathcal{G}(3,0)$:

$$\{1, e_1, e_2, e_3, e_1e_2, e_1e_3, e_2e_3, e_1e_2e_3\}$$

A basis formed this way is called a canonical basis for the geometric algebra, and any other orthogonal basis for V will produce another canonical basis. Each canonical basis consists of 2^n elements. Every multivector of the geometric algebra can be expressed as a linear combination of the canonical basis elements. If the canonical basis elements are $\{B_i \mid i \in S\}$ with S being an index set, then the geometric product of any two multivectors is

$$\left(\sum_i \alpha_i B_i\right)\left(\sum_j \beta_j B_j\right) = \sum_{i,j} \alpha_i \beta_j B_i B_j.$$

The terminology "k-vector" is often encountered to describe multivectors containing elements of only one grade. In higher dimensional space, some such multivectors are not blades (cannot be factored into the outer product of k vectors). By way of example, $e_1 \wedge e_2 + e_3 \wedge e_4$ in R^4 cannot be factored; typically, however, such elements of the algebra do not yield to any geometric interpretation. Only $0, 1, (n-1)$ and n-vectors are always blades in n-space.

Grade Projection

Using an orthogonal basis, a graded vector space structure can be established. Elements of the geometric algebra that are scalar multiples of 1 are grade-0 blades and are called *scalars*. Multivectors that are in the span of $\{e_1,\ldots,e_n\}$ are grade-1 blades and are the ordinary vectors. Multivectors in the span of $\{e_i e_j \mid 1 \leq i < j \leq n\}$ are grade-2 blades and are the bivectors. This terminology continues through to the last grade of n-vectors. Alternatively, grade-n blades are called pseudoscalars, grade-$(n-1)$ blades pseudovectors, etc. Many of the elements of the algebra are not graded by this scheme since they are sums of elements of differing grade. Such elements are said to be of *mixed grade*. The grading of multivectors is independent of the basis chosen originally.

A multivector A may be decomposed with the grade-projection operator $\langle A\rangle_r$, which outputs the grade-r portion of A. As a result:

$$A = \sum_{r=0}^{n} \langle A\rangle_r$$

As an example, the geometric product of two vectors $ab = a\cdot b + a\wedge b = \langle ab\rangle_0 + \langle ab\rangle_2$ since $\langle ab\rangle_0 = a\cdot b$ and $\langle ab\rangle_2 = a\wedge b$ and $\langle ab\rangle_i = 0$ for i other than 0 and 2.

The decomposition of a multivector A may also be split into those components that are even and those that are odd:

$$A^{+} = \langle A\rangle_0 + \langle A\rangle_2 + \langle A\rangle_4 + \cdots$$

$$A^{-} = \langle A\rangle_1 + \langle A\rangle_3 + \langle A\rangle_5 + \cdots$$

This makes the algebra a Z_2-graded algebra or superalgebra with the geometric product. Since the geometric product of two even multivectors is an even multivector, they define an *even subalgebra*. The even subalgebra of an n-dimensional geometric algebra is isomorphic to a full geometric algebra of $(n-1)$ dimensions. Examples include $\mathcal{G}^{+}(2,0) \cong \mathcal{G}(0,1)$ and $\mathcal{G}^{+}(1,3) \cong \mathcal{G}(3,0)$.

Representation of Subspaces

Geometric algebra represents subspaces of V as blades, and so they coexist in the same algebra with vectors from V. A k-dimensional subspace W of V is represented by taking an orthogonal basis $\{b_1, b_2, \ldots, b_k\}$ and using the geometric product to form the blade $D = b_1 b_2 \cdots b_k$. There are multiple blades representing W; all those representing W are scalar multiples of D. These blades can be separated into two sets: positive multiples of D and negative multiples of D. The positive multiples of D are said to have *the same orientation* as D, and the negative multiples the *opposite orientation*.

Blades are important since geometric operations such as projections, rotations and reflections depend on the factorability via the exterior product that (the restricted class of) n-blades provide but that (the generalized class of) grade-n multivectors do not when $n \geq 4$.

Unit Pseudoscalars

Unit pseudoscalars are blades that play important roles in GA. A unit pseudoscalar for a non-degenerate subspace W of V is a blade that is the product of the members of an orthonormal basis for W. It can be shown that if I and I′ are both unit pseudoscalars for W, then $I = \pm I'$ and $I^2 = \pm 1$.

Suppose the geometric algebra $\mathcal{G}(n,0)$ with the familiar positive definite inner product on R^n is formed. Given a plane (2-dimensional subspace) of R^n, one can find an orthonormal basis $\{b_1, b_2\}$ spanning the plane, and thus find a unit pseudoscalar $I = b_1b_2$ representing this plane. The geometric product of any two vectors in the span of b_1 and b_2 lies in $\{\alpha_0 + \alpha_1 I \mid \alpha_i \in R\}$, that is, it is the sum of a 0-vector and a 2-vector.

By the properties of the geometric product, $I^2 = b_1b_2b_1b_2 = -b_1b_2b_2b_1 = -1$. The resemblance to the imaginary unit is not incidental: the subspace $\{\alpha_0 + \alpha_1 I \mid \alpha_i \in \mathbf{R}\}$ is $\mathbf{R}$-algebra isomorphic to the complex numbers. In this way, a copy of the complex numbers is embedded in the geometric algebra for each 2-dimensional subspace of V on which the quadratic form is definite.

It is sometimes possible to identify the presence of an imaginary unit in a physical equation. Such units arise from one of the many quantities in the real algebra that square to -1, and these have geometric significance because of the properties of the algebra and the interaction of its various subspaces.

In $\mathcal{G}(3,0)$, a further familiar case occurs. Given a canonical basis consisting of orthonormal vectors e_i of V, the set of *all* 2-vectors is spanned by

$$\{e_3e_2, e_1e_3, e_2e_1\}.$$

Labelling these i, j and (momentarily deviating from our uppercase convention), the subspace generated by 0-vectors and 2-vectors is exactly $\{\alpha_0 + i\alpha_1 + j\alpha_2 + k\alpha_3 \mid \alpha_i \in R\}$. This set is seen to be a subalgebra, and furthermore is R-algebra isomorphic to the quaternions, another important algebraic system.

Dual Basis

Let $\{e_i\}$ be a basis of V, i.e. a set of linearly independent vectors that span the n-dimensional vector space V. The basis that is dual to $\{e_i\}$ is the set of elements of the dual vector space V^* that forms a biorthogonal system with this basis, thus being the elements denoted $\{e^i\}$ satisfying

$$e^i \cdot e_j = \delta^i{}_j,$$

where δ is the Kronecker delta.

Given a nondegenerate quadratic form on V, V^* becomes naturally identified with V, and the dual basis may be regarded as elements of V, but are not in general the same set as the original basis.

Given further a GA of V, let

$$\epsilon = e_1 \wedge \cdots \wedge e_n$$

be the pseudoscalar (which does not necessarily square to ± 1) formed from the basis $\{e_i\}$. The dual basis vectors may be constructed as

$$e^i = (-1)^{i-1}(e_1 \wedge \cdots \wedge \breve{e}_i \wedge \cdots \wedge e_n)\epsilon^{-1},$$

where the e^i denotes that the ith basis vector is omitted from the product.

Extensions of the Inner and Exterior Products

It is common practice to extend the exterior product on vectors to the entire algebra. This may be done through the use of the grade projection operator:

$$C \wedge D := \sum_{r,s} \langle\langle C\rangle_r \langle D\rangle_s\rangle_{r+s} \quad \text{(the exterior product)}$$

This generalization is consistent with the above definition involving antisymmetrization. Another generalization related to the exterior product is the commutator product:

$$C \times D := \tfrac{1}{2}(CD - DC)$$

The regressive product (usually referred to as the "meet") is the dual of the exterior product (or "join" in this context).[e] The dual specification of elements permits, for blades A and B, the intersection (or meet) where the duality is to be taken relative to the smallest grade blade containing both A and B (the join).

$$C \vee D := ((CI^{-1}) \wedge (DI^{-1}))I$$

with I the unit pseudoscalar of the algebra. The regressive product, like the exterior product, is associative.

The inner product on vectors can also be generalized, but in more than one non-equivalent way. Many authors use the same symbol as for the inner product of vectors for their chosen extension (e.g. Hestenes and Perwass). No consistent notation has emerged.

Among these several different generalizations of the inner product on vectors are:

$$C \rfloor D := \sum_{r,s} \langle\langle C\rangle_r \langle D\rangle_s\rangle_{s-r} \quad \text{(the left contraction)}$$

$$C \lfloor D := \sum_{r,s} \langle\langle C\rangle_r \langle D\rangle_s\rangle_{r-s} \quad \text{(the right contraction)}$$

$$C * D := \sum_{r,s} \langle\langle C\rangle_r \langle D\rangle_s\rangle_0 \quad \text{(the scalar product)}$$

$$C \bullet D := \sum_{r,s} \langle\langle C\rangle_r \langle D\rangle_s\rangle_{|s-r|} \quad \text{(the "(fat) dot" product)}$$

A number of identities incorporating the contractions are valid without restriction of their inputs.

For example,

$$C \rfloor D = (C \wedge (DI^{-1}))I$$
$$C \lfloor D = I((I^{-1}C) \wedge D)$$
$$(A \wedge B) * C = A * (B \rfloor C)$$
$$C * (B \wedge A) = (C \lfloor B) * A$$
$$A \rfloor (B \rfloor C) = (A \wedge B) \rfloor C.$$

Benefits of using the left contraction as an extension of the inner product on vectors include that the identity $ab = a \cdot b + a \wedge b$ is extended to $aB = a \rfloor B + a \wedge B$ for any vector a and multivector B, and that the projection operation $\mathcal{P}_b(a) = (a \cdot b^{-1})b$ is extended to $\mathcal{P}_B(A) = (A \rfloor B^{-1}) \rfloor B$ for any blade B and any multivector A (with a minor modification to accommodate null B, given below).

Linear Functions

Although versors are inherently more useful, they are a subgroup of linear functions on multivectors and these can still be used when necessary. The geometric algebra of an n-dimensional vector space is spanned by a basis of 2^n elements. If a multivector is represented by a $2^n \times 1$ real column matrix in terms of a basis, then all linear transformations of the multivector can be expressed as the matrix multiplication by a $2^n \times 2^n$ real matrix. However, such a general linear transformation allows arbitrary exchanges among grades, such as a "rotation" of a scalar into a vector, which has no evident geometric interpretation.

A general linear transformation from vectors to vectors is of interest. With the natural restriction to preserving the induced exterior algebra, the *outermorphism* of the linear transformation is its unique extension. If f is a linear function that maps vectors to vectors, then its outermorphism is the function that obeys the rule

$$\underline{f}(a_1 \wedge a_2 \wedge \cdots \wedge a_r) = f(a_1) \wedge f(a_2) \wedge \cdots \wedge f(a_r)$$

for a blade, extended to the whole algebra through linearity.

Modeling Geometries

Although a lot of attention has been placed on CGA in recent years, it is to be noted that GA is not just one algebra, it is one of a family of algebras with the same essential structure.

Vector Space Model

$\mathcal{G}(3,0)$ may be considered as an extension or completion of vector algebra. *From Vectors to Geometric Algebra* covers basic analytic geometry and gives an introduction to stereographic projection.

The even subalgebra of $\mathcal{G}(2,0)$ is isomorphic to the complex numbers, as may be seen by writing a vector P in terms of its components in an orthonormal basis and left multiplying by the basis vector e_1, yielding

$$Z = e_1 P = e_1(xe_1 + ye_2) = x(1) + y(e_1 e_2),$$

where we identify $\mathrm{i} \mapsto -\mathrm{e}_1\mathrm{e}_2$ since

$$(e_1 e_2)^2 = e_1 e_2 e_1 e_2 = -e_1 e_1 e_2 e_2 = -1 .$$

Similarly, the even subalgebra of $\mathcal{G}(3,0)$ with basis $\{1, e_2 e_3, e_3 e_1, e_1 e_2\}$ is isomorphic to the quaternions as may be seen by identifying $\mathrm{i} \mapsto -\mathrm{e}_2\mathrm{e}_3$,, $\mathrm{j} \mapsto -\mathrm{e}_3\mathrm{e}_1$ and $\mathrm{k} \mapsto -\mathrm{e}_1\mathrm{e}_2$.

Every associative algebra has a matrix representation; replacing the three Cartesian basis vectors by the Pauli matrices gives a representation of $\mathcal{G}(3,0)$:

$$e_1 = \sigma_1 = \sigma_x = \begin{pmatrix} 0 & 1 \\ 1 & 0 \end{pmatrix}$$

$$e_2 = \sigma_2 = \sigma_y = \begin{pmatrix} 0 & -i \\ i & 0 \end{pmatrix}$$

$$e_3 = \sigma_3 = \sigma_z = \begin{pmatrix} 1 & 0 \\ 0 & -1 \end{pmatrix}$$

Dotting the "Pauli vector" (a dyad):

$\sigma = \sigma_1 e_1 + \sigma_2 e_2 + \sigma_3 e_3$ with arbitrary vectors a and b and multiplying through gives:

$(\sigma \cdot a)(\sigma \cdot b) = a \cdot b + a \wedge b$ (Equivalently, by inspection, $a \cdot b + i\sigma \cdot (a \times b)$)

Spacetime Model

In physics, the main applications are the geometric algebra of Minkowski 3+1 spacetime, $G(1,3)$, called spacetime algebra (STA), or less commonly, $G(1,3)$ (R), called the algebra of physical space (APS), where $G(1,3)$ (R) is isomorphic to the *even* subalgebra of the 3+1 algebra, $G^{+}_{1,3}$ (R).

While in STA points of spacetime are represented simply by vectors, in APS, points of $(3+1)$-dimensional spacetime are instead represented by paravectors: a 3-dimensional vector (space) plus a 1-dimensional scalar (time).

In spacetime algebra the electromagnetic field tensor has a bivector representation $F = (E + icB)\gamma_0$. Here, the $i = \gamma_0\gamma_1\gamma_2\gamma_3$ is the unit pseudoscalar (or four-dimensional volume element), γ_0 is the unit vector in time direction, and E and B are the classic electric and magnetic field vectors (with a zero time component). Using the four-current J, Maxwell's equations then become

Formulation	Homogeneous equations	Non-homogeneous equations
Fields	$DF = \mu_0 J$	
	$D \wedge F = 0$	$D \cdot F = \mu_0 J$
Potentials (any gauge)	$F = D \wedge A$	$D \cdot D \wedge A = \mu_0 J$
Potentials (Lorenz gauge)	$F = DA$ $D \cdot A = 0$	$D^2 A = \mu_0 J$

In geometric calculus, juxtapositioning of vectors such as in DF indicate the geometric product and can be decomposed into parts as $DF = D \cdot F + D \wedge F$. Here D is the covector derivative in any spacetime and reduces to ∇ in flat spacetime. Where $\triangledown$ plays a role in Minkowski 4-spacetime which is synonymous to the role of ∇ in Euclidean 3-space and is related to the d'Alembertian by $\Box = \triangledown^2$. Indeed, given an observer represented by a future pointing timelike vector γ_0 we have

$$\gamma_0 \cdot \triangledown = \frac{1}{c}\frac{\partial}{\partial t}$$

$$\gamma_0 \wedge \triangledown = \nabla$$

Boosts in this Lorentzian metric space have the same expression e^{β} as rotation in Euclidean space, where β is the bivector generated by the time and the space directions involved, whereas in the Euclidean case it is the bivector generated by the two space directions, strengthening the "analogy" to almost identity.

The Dirac matrices are a representation of $\mathcal{G}(1,3)$, showing the equivalence with matrix representations used by physicists.

Homogeneous Model

The first model here is $\mathcal{G}(4,0)$, the GA version of homogeneous coordinates used in projective geometry. Here a vector represents a point and an outer product of vectors an oriented length yet we may work with the algebra in just the same way as in $\mathcal{G}(3,0)$. However, a useful inner product cannot be defined in the space and so there is no geometric product either leaving only outer product and non-metric uses of duality such as meet and join.

Nevertheless, there has been investigation of 4-dimensional alternatives to the full 5-dimensional CGA for limited geometries such as rigid body movements. A selection of these can be found in Part IV of Guide to Geometric Algebra in Practice . Note that the algebra $\mathcal{G}(3,0,1)$ appears as a subalgebra of CGA by selecting just one null vector and dropping the other and further that the "motor algebra" (isomorphic to dual quaternions) is the even subalgebra of $\mathcal{G}(3,0,1)$.

Conformal Model

Working within GA, Euclidean space $\mathcal{E}^3$ is embedded projectively in the CGA $\mathcal{G}^{4,1}$ via the identification of Euclidean points with 1-d subspaces in the 4-d null cone of the 5-d CGA vector subspace, and adding a point at infinity. This allows all conformal transformations to be done as rotations and reflections and is covariant, extending incidence relations of projective geometry to circles and spheres.

Specifically, we add orthogonal basis vectors e_+ and e_- such that ${e_+}^2 = +1$ and ${e_-}^2 = -1$ to the basis of $\mathcal{G}(3,0)$ and identify null vectors

$n_\infty = e_- + e_+$ as an ideal point (point at infinity) and

$n_o = \frac{1}{2}(e_- - e_+)$ as the point at the origin, giving

$n_\infty \cdot n_o = -1$.

This procedure has some similarities to the procedure for working with homogeneous coordinates in projective geometry and in this case allows the modeling of Euclidean transformations as orthogonal transformations.

A fast changing and fluid area of GA, CGA is also being investigated for applications to relativistic physics.

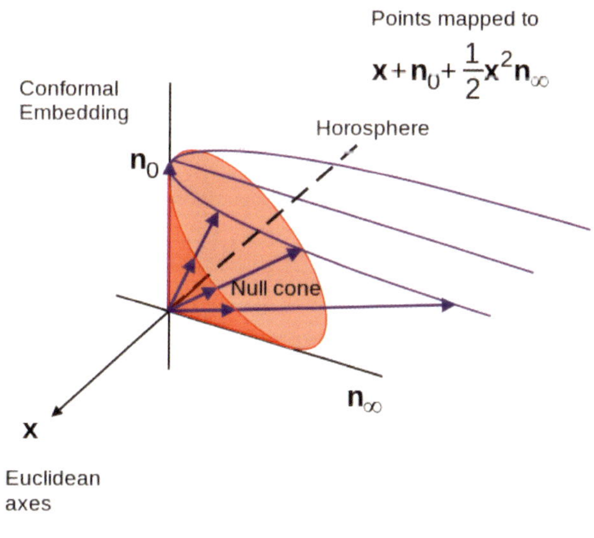

Geometric Interpretation

Projection and Rejection

For any vector a and any invertible vector m,

$$a = amm^{-1} = (a \cdot m + a \wedge m)m^{-1} = a_{\|m} + a_{\perp m},$$

where the projection of a onto m (or the parallel part) is

$$a_{\parallel m} = (a \cdot m)m^{-1}$$

and the rejection of a from m (or the orthogonal part) is

$$a_{\perp m} = a - a_{\parallel m} = (a \wedge m)m^{-1}.$$

Using the concept of a k-blade B as representing a subspace of V and every multivector ultimately being expressed in terms of vectors, this generalizes to projection of a general multivector onto any invertible k-blade B as

$$\mathcal{P}_{\mathrm{B}}(\mathrm{A}) = (\mathrm{A} \rfloor \mathrm{B}^{-1}) \rfloor \mathrm{B},$$

with the rejection being defined as

$$\mathcal{P}_B^{\perp}(A) = A - \mathcal{P}_B(A).$$

The projection and rejection generalize to null blades B by replacing the inverse B^{-1} with the pseudoinverse B^{+} with respect to the contractive product. The outcome of the projection coincides in both cases for non-null blades. For null blades B, the definition of the projection given here with the first contraction rather than the second being onto the pseudoinverse should be used, as only then is the result necessarily in the subspace represented by B. The projection generalizes through linearity to general multivectors A. The projection is not linear in B and does not generalize to objects B that are not blades.

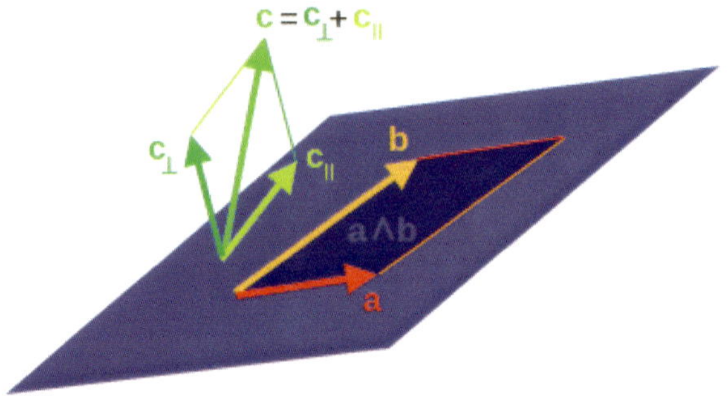

In 3-d space, a bivector $a \wedge b$ defines a 2-d plane subspace (light blue, extends infinitely in indicated directions). Any vector c in 3-d space can be decomposed into its projection $c_{\parallel}$ onto a plane and its rejection $c_{\perp}$ from this plane.

Reflection

Simple reflections in a hyperplane are readily expressed in the algebra through conjugation with a single vector. These serve to generate the group of general rotoreflections and rotations.

The reflection c' of a vector c along a vector m, or equivalently in the hyperplane orthogonal to m, is the same as negating the component of a vector parallel to m. The result of the reflection will be

$$c' = -c_{\parallel m} + c_{\perp m} = -(c \cdot m)m^{-1} + (c \wedge m)m^{-1} = (-m \cdot c - m \wedge c)m^{-1} = -mcm^{-1}$$

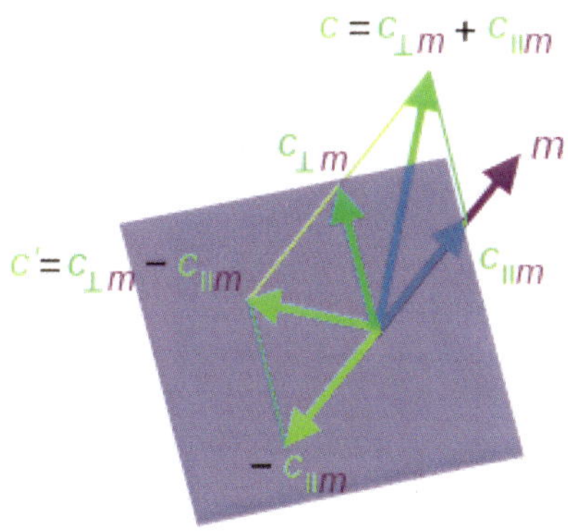

Reflection of vector c along a vector m. Only the component of c parallel to m is negated.

This is not the most general operation that may be regarded as a reflection when the dimension $n \geq 4$. A general reflection may be expressed as the composite of any odd number of single-axis reflections. Thus, a general reflection a' of a vector a may be written

$$a \mapsto a' = -MaM^{-1},$$

where

$$M = pq \cdots r \text{ and } M^{-1} = (pq \cdots r)^{-1} = r^{-1} \cdots q^{-1}p^{-1}.$$

If we define the reflection along a non-null vector m of the product of vectors as the reflection of every vector in the product along the same vector, we get for any product of an odd number of vectors that, by way of example,

$$(abc)' = a'b'c' = (-mam^{-1})(-mbm^{-1})(-mcm^{-1}) = -ma(m^{-1}m)b(m^{-1}m)cm^{-1} = -mabcm^{-1}$$

and for the product of an even number of vectors that

$$(abcd)' = a'b'c'd' = (-mam^{-1})(-mbm^{-1})(-mcm^{-1})(-mdm^{-1}) = mabcdm^{-1}$$

Using the concept of every multivector ultimately being expressed in terms of vectors, the reflection of a general multivector A using any reflection versor M may be written

$$A \mapsto M\alpha(A)M^{-1},$$

where α is the automorphism of reflection through the origin of the vector space $(v \mapsto -v)$ extended through linearity to the whole algebra.

Rotations

If we have a product of vectors $R = a_1 a_2 \cdots a_r$ then we denote the reverse as

$$R^{\dagger} = (a_1 a_2 \cdots a_r)^{\dagger} = a_r \cdots a_2 a_1.$$

As an example, assume that $R = ab$ we get

$$RR^{\dagger} = abba = ab^2a = a^2b^2 = ba^2b = baab = R^{\dagger}R.$$

Scaling R so that $RR^{\dagger} = 1$ then

$$(RvR^{\dagger})^2 = Rv^2R^{\dagger} = v^2RR^{\dagger} = v^2$$

so $RvR^{\dagger}$ leaves the length of v unchanged. We can also show that

$$(Rv_1R^{\dagger}) \cdot (Rv_2R^{\dagger}) = v_1 \cdot v_2$$

so the transformation $RvR^{\dagger}$ preserves both length and angle. It therefore can be identified as a rotation or rotoreflection; R is called a rotor if it is a proper rotation (as it is if it can be expressed as a product of an even number of vectors) and is an instance of what is known in GA as a *versor*.

There is a general method for rotating a vector involving the formation of a multivector of the form $R = e^{-B\theta/2}$ that produces a rotation θ in the plane and with the orientation defined by a 2-blade B.

Rotors are a generalization of quaternions to n-dimensional spaces.

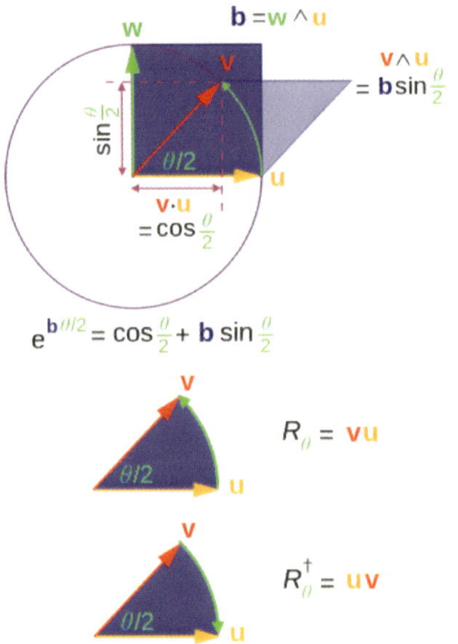

A rotor that rotates vectors in a plane rotates vectors through angle θ, that is $x \mapsto R_\theta x R_\theta^{\dagger}$ is a rotation of x through angle θ. The angle between u and v is $\theta / 2$. Similar interpretations are valid for a general multivector X instead of the vector x.

Versor

A k-versor is a multivector represented by the geometric product of k invertible vectors. Unit quaternions (originally called versors by Hamilton) may be identified with rotors in 3D space in much the same way as real 2D rotors subsume complex numbers.

Some authors use the term "versor product" to refer to the frequently occurring case where an operand is "sandwiched" between operators. The descriptions for rotations and reflections, including their outermorphisms, are examples of such sandwiching. The outermorphisms have a particularly simple algebraic form. Specifically, a mapping of vectors of the form

$$\mathrm{V} \to \mathrm{V} : \mathrm{a} \mapsto \mathrm{RaR}^{-1} \text{ extends to the outermorphism } \mathcal{G}(\mathrm{V}) \to \mathcal{G}(\mathrm{V}) : \mathrm{A} \mapsto \mathrm{RAR}^{-1}.$$

Since both operators and operand are versors there is potential for alternative examples such as rotating a rotor or reflecting a spinor always provided that some geometrical or physical significance can be attached to such operations.

By the Cartan–Dieudonné theorem we have that every isometry can be given as reflections in hyperplanes and since composed reflections provide rotations then we have that orthogonal transformations are versors.

In group terms, for a real, non-degenerate $\mathcal{G}(p,q)$, having identified the group $\mathcal{G}^X$ as the group of all invertible elements of $\mathcal{G}$, Lundholm gives a proof that the "versor group" $\{v_1v_2....v_k \in G : V_i \in V^X$ (set of invertible vectors)} is equal to the Lipschitz group Γ (aka Clifford group, although Lundholm deprecates this usage).

Subgroups of Γ

Lundholm defines the Pin, Spin, and Spin+ subgroups, generated by unit vectors, and in the case of Spin and Spin+, only an even number of such vector factors can be present .

Subgroup	Definition	Description
Pin	$x \in \Gamma : xx^{\dagger} = \pm 1$	unit versors
$Spin$	$Pin \cap \mathcal{G}^{+}$	even unit versors
$Spin^{+}$	$x \in Spin : xx^{\dagger} = 1$	rotors

Spinors are defined as elements of the even subalgebra of a real GA; an analysis of the GA approach to spinors is given by Francis and Kosowsky .

Examples and applications

Hypervolume of a parallelotope spanned by vectors

For vectors a and b spanning a parallelogram we have

$$a \wedge b = ((a \wedge b)b^{-1})b = a_{\perp b}b$$

with the result that $a \wedge b$ is linear in the product of the "altitude" and the "base" of the parallelogram, that is, its area.

Similar interpretations are true for any number of vectors spanning an n-dimensional parallelotope; the exterior product of vectors $a_1, a_2, \ldots, a_n$, that is $\bigwedge_{i=1}^{n} a_i$, has a magnitude equal to the volume of the n-parallelotope. An n-vector does not necessarily have a shape of a parallelotope – this is a convenient visualization. It could be any shape, although the volume equals that of the parallelotope.

Intersection of a Line and a Plane

We may define the line parametrically by $p = t + \alpha\ v$ where p and t are position vectors for points T and P and v is the direction vector for the line.

Then

$$B \wedge (p-q) = 0 \text{ and } B \wedge (t+\alpha v-q) = 0$$

so

$$\alpha = \frac{B \wedge (q-t)}{B \wedge v}$$

and

$$p = t + \left(\frac{B \wedge (q-t)}{B \wedge v} \right) v.$$

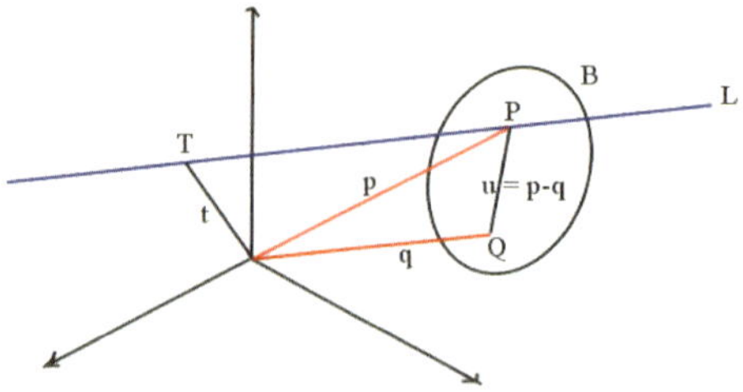

A line L defined by points T and P (which we seek) and a plane defined by a bivector B containing points P and Q.

Rotating Systems

The mathematical description of rotational forces such as torque and angular momentum often makes use of the cross product of vector calculus in three dimensions with a convention of orientation (handedness).

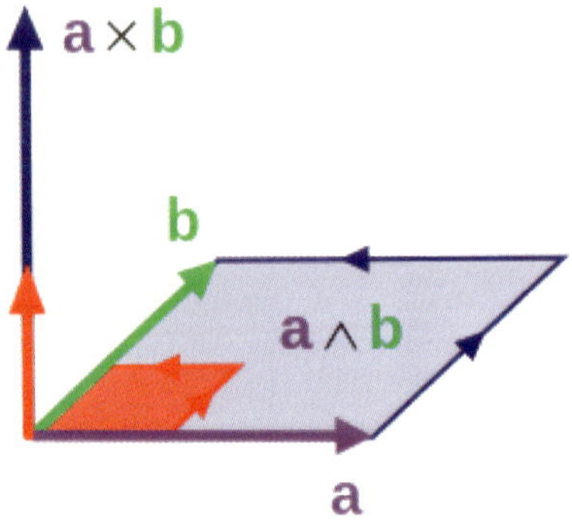

The cross product in relation to the exterior product. In red are the unit normal vector, and the "parallel" unit bivector.

The cross product can be viewed in terms of the exterior product allowing a more natural geometric interpretation of the cross product as a bivector using the dual relationship $a \times b = -I(a \wedge b)$.

For example, torque is generally defined as the magnitude of the perpendicular force component times distance, or work per unit angle.

Suppose a circular path in an arbitrary plane containing orthonormal vectors $\hat{u}$ and $\hat{v}$ is parameterized by angle.

$$r = r(\hat{u}\cos\theta + \hat{v}\sin\theta) = r\hat{u}(\cos\theta + \hat{u}\hat{v}\sin\theta)$$

By designating the unit bivector of this plane as the imaginary number

$$i = \hat{u}\hat{v} = \hat{u} \wedge \hat{v}$$

$$i^2 = -1$$

this path vector can be conveniently written in complex exponential form

$$r = r\hat{u}e^{i\theta}$$

and the derivative with respect to angle is

$$\frac{dr}{d\theta} = r\hat{u}ie^{i\theta} = ri.$$

So the torque, the rate of change of work W, due to a force F, is

$$\tau = \frac{dW}{d\theta} = F\cdot\frac{dr}{d\theta} = F\cdot(ri).$$

Unlike the cross product description of torque, $\tau = r \times F$, , the geometric algebra description does not introduce a vector in the normal direction; a vector that does not exist in two and that is not unique in greater than three dimensions. The unit bivector describes the plane and the orientation of the rotation, and the sense of the rotation is relative to the angle between the vectors $\hat{u}$ and $\hat{v}$.

Geometric Calculus

Geometric calculus extends the formalism to include differentiation and integration including differential geometry and differential forms.

Essentially, the vector derivative is defined so that the GA version of Green's theorem is true,

$$\int_A dA\nabla f = \oint_{\partial A} dx f$$

and then one can write

$$\nabla f = \nabla\cdot f + \nabla \wedge f$$

as a geometric product, effectively generalizing Stokes' theorem (including the differential form version of it).

In $1D$ when A is a curve with endpoints a and b, then

$$\int_A dA\nabla f = \oint_{\partial A} dx f$$

reduces to

$$\int_a^b dx \nabla f = \int_a^b dx . \nabla f = \int_a^b df = f(b) - f(a)$$

or the fundamental theorem of integral calculus.

Also developed are the concept of vector manifold and geometric integration theory (which generalizes Cartan's differential forms).

Vector Algebra

In mathematics and linear algebra, vector algebra refers to algebraic operations in vector spaces. Most commonly, it refers to operations on Euclidean vectors.

General Vector Spaces

Addition and Subtraction

Assume now that a and b are not necessarily equal vectors, but that they may have different magnitudes and directions. The sum of a and b is defined as:

$$\mathbf{a} + \mathbf{b} = \sum_{i=1}^{n} (a_i + b_i)\mathbf{e}_i = (a_1 + b_1)\mathbf{e}_1 + (a_2 + b_2)\mathbf{e}_2 + \cdots + (a_n + b_n)\mathbf{e}_n$$

The addition may be represented graphically by placing the tail of the arrow b at the head of the arrow a, and then drawing an arrow from the tail of a to the head of b. The new arrow drawn represents the vector a + b, as illustrated below:

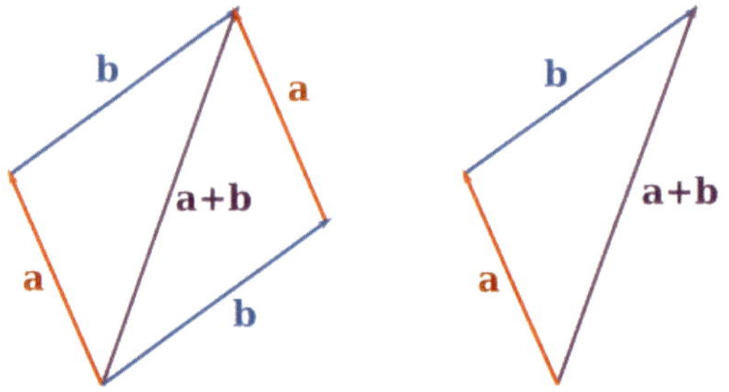

This addition method is sometimes called the *parallelogram rule* because a and b form the sides of a parallelogram and a + b is one of the diagonals. If a and b are bound vectors that have the same base point, this point will also be the base point of a + b. One can check geometrically that a + b = b + a and (a + b) + c = a + (b + c).

The difference of a and b is

$$\mathbf{a} - \mathbf{b} = \sum_{i=1}^{n} (a_i - b_i)\mathbf{e}_i = (a_1 - b_1)\mathbf{e}_1 + (a_2 - b_2)\mathbf{e}_2 + \cdots + (a_n - b_n)\mathbf{e}_n$$

Subtraction of two vectors can be geometrically defined as follows: to subtract b from a, place the tails of a and b at the same point, and then draw an arrow from the head of b to the head of a. This new arrow represents the vector a – b, as illustrated below:

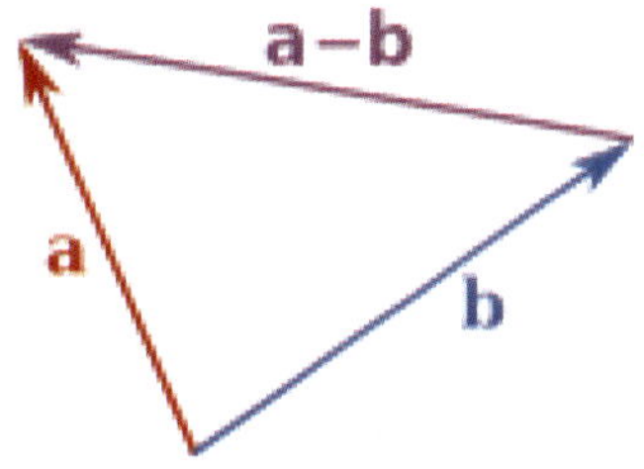

Subtraction of two vectors may also be performed by adding the opposite of the second vector to the first vector, that is, a – b = a + (–b).

Scalar Multiplication

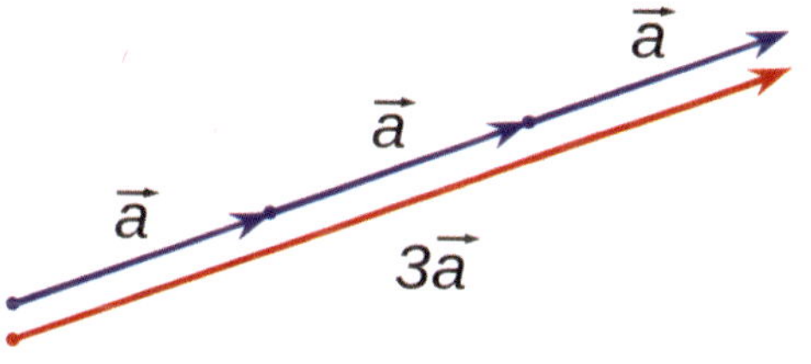

Scalar multiplication of a vector by a factor of 3 stretches the vector out.

A vector may also be multiplied, or re-*scaled*, by a real number *r*. In the context of conventional vector algebra, real numbers are often called scalars (from *scale*) to distinguish them from vectors. The operation of multiplying a vector by a scalar is called *scalar multiplication*. The resulting vector is

$$\mathbf{ra} = \sum_{i=1}^{n} (ra_i)\mathbf{e}_i = (ra_1)\mathbf{e}_1 + (ra_2)\mathbf{e}_2 + \cdots + (ra_n)\mathbf{e}_n .$$

Intuitively, multiplying by a scalar *r* stretches a vector out by a factor of *r*. Geometrically, this can be visualized (at least in the case when *r* is an integer) as placing *r* copies of the vector in a line where the endpoint of one vector is the initial point of the next vector.

If *r* is negative, then the vector changes direction: it flips around by an angle of 180°. Two examples ($r = -1$ and $r = 2$) are given below:

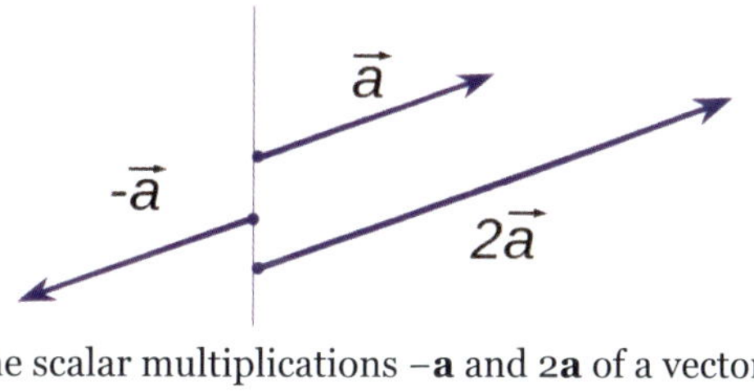

The scalar multiplications −**a** and 2**a** of a vector **a**

Scalar multiplication is distributive over vector addition in the following sense: *r*(a + b) = *r*a+ *r*b for all vectors a and b and all scalars *r*. One can also show that a – b = a + (–1)b.

Dot Product

The *dot product* of two vectors a and b (sometimes called the *inner product*, or, since its result is a scalar, the *scalar product*) is denoted by a · b and is defined as:

$$\mathbf{a} \cdot \mathbf{b} = \|\mathbf{a}\|\|\mathbf{b}\|\cos\theta$$

where θ is the measure of the angle between a and b. Geometrically, this means that a and b are drawn with a common start point and then the length of a is multiplied with the length of that component of b that points in the same direction as a.

The dot product can also be defined as the sum of the products of the components of each vector as

$$\mathbf{a}\cdot\mathbf{b} = \sum_{i=1}^{n} a_i b_i = a_1 b_1 + a_2 b_2 + \cdots + a_n b_n$$

Three Dimensional Space

Cross Product

The *cross product* (also called the *vector product* or *outer product*) is only meaningful in three or seven dimensions. The cross product differs from the dot product primarily in that the result of the cross product of two vectors is a vector. The cross product, denoted a × b, is a vector perpendicular to both a and b and is defined as

$$\mathbf{a} \times \mathbf{b} = \|\mathbf{a}\|\|\mathbf{b}\|\sin(\theta)\mathbf{n}$$

where θ is the measure of the angle between a and b, and n is a unit vector perpendicular to both a and b which completes a right-handed system. The right-handedness constraint is necessary because there exist *two* unit vectors that are perpendicular to both a and b, namely, n and (−n).

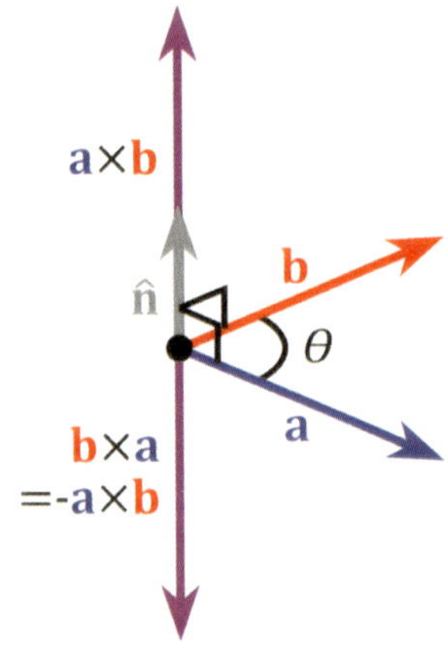

An illustration of the cross product

The cross product a × b is defined so that a, b, and a × b also becomes a right-handed system (but note that a and b are not necessarily orthogonal). This is the right-hand rule.

The length of a × b can be interpreted as the area of the parallelogram having a and b as sides.

The cross product can be written as

$$\mathbf{a} \times \mathbf{b} = (a_2 b_3 - a_3 b_2)\mathbf{e}_1 + (a_3 b_1 - a_1 b_3)\mathbf{e}_2 + (a_1 b_2 - a_2 b_1)\mathbf{e}_3 .$$

For arbitrary choices of spatial orientation (that is, allowing for left-handed as well as right-handed coordinate systems) the cross product of two vectors is a pseudovector instead of a vector.

Other Dimensions

The cross product does not readily generalise to other dimensions, though the closely related exterior product does, whose result is a bivector. In two dimensions this is simply a pseudoscalar

$$(a_1 \mathbf{e}_1 + a_2 \mathbf{e}_2) \wedge (b_1 \mathbf{e}_1 + b_2 \mathbf{e}_2) = (a_1 b_2 - a_2 b_1)\mathbf{e}_1 \wedge \mathbf{e}_2 .$$

A seven-dimensional cross product is similar to the cross product in that its result is a vector orthogonal to the two arguments; there is however no natural way of selecting one of the possible such products.

Comparison of Vector Algebra and Geometric Algebra

Vector algebra and geometric algebra are complementary approaches to providing additional algebraic structures on vector spaces, with geometric interpretations, particularly vector fields in multivariable calculus and applications in mathematical physics.

Vector algebra is specific to Euclidean 3-space, while geometric algebra uses multilinear algebra and applies in all dimensions and signatures, notably 3+1 spacetime as well as 2 dimensions. They are mathematically equivalent in 3 dimensions, where this article is focused, though the approaches differ. Vector algebra is more widely used in elementary multivariable calculus, while geometric algebra is used in more advanced treatments, and is proposed for elementary use as well.

Basic Concepts and Operations

Geometric algebra (GA) is an extension or completion of vector algebra (VA). The reader is herein assumed to be familiar with the basic concepts and operations of VA and this article will mainly concern itself with operations in $\mathcal{G}_3$ the GA of 3D space (nor is this article intended to be mathematically rigorous). In GA, vectors are not normally written boldface as the meaning is usually clear from the context.

The fundamental difference is that GA provides an additional product of vectors called the "geometric product". Elements of GA are graded multivectors, scalars are grade 0, the usual vectors are grade 1, bivectors are grade 2 and the highest grade (3 in the 3D case) is traditionally called the pseudoscalar and designated I.

The ungeneralized 3D vector form of the geometric product is :

$$ab = a \cdot b + a \wedge b$$

that is the sum of the usual dot (inner) product and the outer (exterior) product (this last is closely related to the cross product and will be explained below).

In VA, entities such as pseudovectors and pseudoscalars need to be bolted on, whereas in GA the equivalent bivector and pseudovector respectively exist naturally as subspaces of the algebra.

For example, applying vector calculus in 2 dimensions, such as to compute torque or curl, requires adding an artificial 3rd dimension and extending the vector field to be constant in that dimension, or alternately considering these to be scalars. The torque or curl is then a normal vector field in this 3rd dimension. By contrast, geometric algebra in 2 dimensions defines these as a pseudoscalar field (a bivector), without requiring a 3rd dimension. Similarly, the scalar triple product is ad hoc, and can instead be expressed uniformly using the exterior product and the geometric product.

Translations between Formalisms

Here are some comparisons between standard $\mathbb{R}^3$ vector relations and their corresponding exterior product and geometric product equivalents. All the exterior and geometric product equivalents here are good for more than three dimensions, and some also for two. In two dimensions the cross product is undefined even if what it describes (like torque) is perfectly well defined in a plane without introducing an arbitrary normal vector outside of the space.

Many of these relationships only require the introduction of the exterior product to generalize, but since that may not be familiar to somebody with only a background in vector algebra and calculus, some examples are given.

Cross and Exterior Products

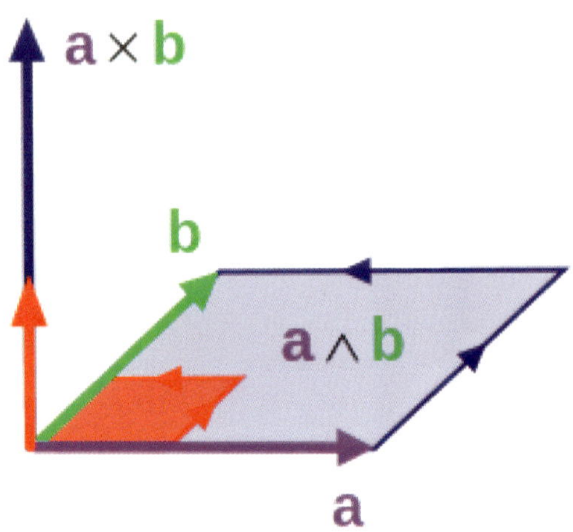

The cross product in relation to the exterior product. In red are the orthogonal unit vector, and the "parallel" unit bivector.

$\mathbf{u} \times \mathbf{v}$ is perpendicular to the plane containing $\mathbf{u}$ and $\mathbf{v}$.

$\mathbf{u} \wedge \mathbf{v}$ is an oriented representation of the same plane.

We have the pseudoscalar $I = e_1 e_2 e_3$ (right handed orthonormal frame) and so

$e_1 I = I e_1 = e_2 e_3$ returns a bivector and

$I(e_2 \wedge e_3) = I e_2 e_3 = -e_1$ returns a vector perpendicular to the $e_2 \wedge e_3$ plane.

This yields a convenient definition for the cross product of traditional vector algebra:

$$u \times v = -I(u \wedge v)$$

(this is antisymmetric). Relevant is the distinction between axial and polar vectors in vector algebra, which is natural in geometric algebra as the distinction between vectors and bivectors (elements of grade two).

The i here is a unit pseudoscalar of Euclidean 3-space, which establishes a duality between the vectors and the bivectors, and is named so because of the expected property

$$i^2 = (e_1 e_2 e_3)^2 = e_1 e_2 e_3 e_1 e_2 e_3 = -e_1 e_2 e_1 e_3 e_2 e_3 = e_1 e_1 e_2 e_3 e_2 e_3 = -e_3 e_2 e_2 e_3 = -1$$

The equivalence of the $\mathbb{R}^3$ cross product and the exterior product expression above can be confirmed by direct multiplication of $-i = -e_1 e_2 e_3$ with a determinant expansion of the exterior product

$$u \wedge v = \sum_{1 \le i < j \le 3} (u_i v_j - v_i u_j) e_i \wedge e_j = \sum_{1 \le i < j \le 3} (u_i v_j - v_i u_j) e_i e_j$$

Cross product as an exterior product. Essentially, the geometric product of a bivector and the pseudoscalar of Euclidean 3-space provides a method of calculation of the Hodge dual.

Norm of a Vector

Ordinarily:

$$\| \mathbf{u} \|^2 = \mathbf{u} \cdot \mathbf{u}$$

Making use of the geometric product and the fact that the exterior product of a vector wedge with itself is zero:

$$\mathbf{uu} = \| \mathbf{u} \|^2 = \mathbf{u}^2 = \mathbf{u} \cdot \mathbf{u} + \mathbf{u} \wedge \mathbf{u} = \mathbf{u} \cdot \mathbf{u}$$

Lagrange Identity

In three dimensions the product of two vector lengths can be expressed in terms of the dot and cross products

$$\| \mathbf{u} \|^2 \| \mathbf{v} \|^2 = (\mathbf{u} \cdot \mathbf{v})^2 + \| \mathbf{u} \times \mathbf{v} \|^2$$

The corresponding generalization expressed using the geometric product is

$$\| \mathbf{u} \|^2 \| \mathbf{v} \|^2 = (\mathbf{u} \cdot \mathbf{v})^2 - (\mathbf{u} \wedge \mathbf{v})^2$$

This follows from expanding the geometric product of a pair of vectors with its reverse

$$(\mathbf{uv})(\mathbf{vu}) = (\mathbf{u} \cdot \mathbf{v} + \mathbf{u} \wedge \mathbf{v})(\mathbf{u} \cdot \mathbf{v} - \mathbf{u} \wedge \mathbf{v})$$

Determinant expansion of cross and wedge products

$$\mathbf{u} \times \mathbf{v} = \sum_{i<j} \begin{vmatrix} u_i & u_j \\ v_i & v_j \end{vmatrix} \mathbf{e}_i \times \mathbf{e}_j$$

$$\mathbf{u} \wedge \mathbf{v} = \sum_{i<j} \begin{vmatrix} u_i & u_j \\ v_i & v_j \end{vmatrix} \mathbf{e}_i \wedge \mathbf{e}_j$$

Without justification or historical context, traditional linear algebra texts will often define the determinant as the first step of an elaborate sequence of definitions and theorems leading up to the solution of linear systems, Cramer's rule and matrix inversion.

An alternative treatment is to axiomatically introduce the wedge product, and then demonstrate that this can be used directly to solve linear systems. This is shown below, and does not require sophisticated math skills to understand.

It is then possible to define determinants as nothing more than the coefficients of the wedge product in terms of "unit *k*-vectors" ($\mathbf{e}_i \wedge \mathbf{e}_j$ terms) expansions as above.

A one-by-one determinant is the coefficient of $\mathbf{e}_1$ for an $\mathbb{R}^1$ 1-vector.

A two-by-two determinant is the coefficient of $\mathbf{e}_1 \wedge \mathbf{e}_2$ for an $\mathbb{R}^2$ bivector

A three-by-three determinant is the coefficient of $\mathbf{e}_1 \wedge \mathbf{e}_2 \wedge \mathbf{e}_3$ for an $\mathbb{R}^3$ trivector

...

When linear system solution is introduced via the wedge product, Cramer's rule follows as a side-effect, and there is no need to lead up to the end results with definitions of minors, matrices, matrix invertibility, adjoints, cofactors, Laplace expansions, theorems on determinant multiplication and row column exchanges, and so forth.

Matrix Related

Matrix inversion (Cramer's rule) and determinants can be naturally expressed in terms of the wedge product.

The use of the wedge product in the solution of linear equations can be quite useful for various geometric product calculations.

Traditionally, instead of using the wedge product, Cramer's rule is usually presented as a generic algorithm that can be used to solve linear equations of the form $Ax = b$ (or equivalently to invert a matrix). Namely

$$x = \frac{1}{|A|} \operatorname{adj}(A) b .$$

This is a useful theoretic result. For numerical problems row reduction with pivots and other methods are more stable and efficient.

When the wedge product is coupled with the Clifford product and put into a natural geometric context, the fact that the determinants are used in the expression of $\mathbb{R}^N$ parallelogram area and parallelepiped volumes (and higher-dimensional generalizations thereof) also comes as a nice side-effect.

As is also shown below, results such as Cramer's rule also follow directly from the wedge product's selection of non-identical elements. The end result is then simple enough that it could be derived easily if required instead of having to remember or look up a rule.

Two variables example

$$\begin{bmatrix} a & b \end{bmatrix} \begin{bmatrix} x \\ y \end{bmatrix} = ax + by = c .$$

Pre- and post-multiplying by a and b,

$$(ax + by) \wedge b = (a \wedge b)x = c \wedge b$$

$$a \wedge (ax + by) = (a \wedge b)y = a \wedge c$$

Provided $a \wedge b \neq 0$ the solution is

$$\begin{bmatrix} x \\ y \end{bmatrix} = \frac{1}{a \wedge b} \begin{bmatrix} c \wedge b \\ a \wedge c \end{bmatrix}.$$

For $a, b \in \mathbb{R}^2$, this is Cramer's rule since the $e_1 \wedge e_2$ factors of the wedge products

$$u \wedge v = \begin{vmatrix} u_1 & u_2 \\ v_1 & v_2 \end{vmatrix} e_1 \wedge e_2$$

divide out.

Similarly, for three, or N variables, the same ideas hold

$$\begin{bmatrix} a & b & c \end{bmatrix} \begin{bmatrix} x \\ y \\ z \end{bmatrix} = d$$

$$\begin{bmatrix} x \\ y \\ z \end{bmatrix} = \frac{1}{a \wedge b \wedge c} \begin{bmatrix} d \wedge b \wedge c \\ a \wedge d \wedge c \\ a \wedge b \wedge d \end{bmatrix}$$

Again, for the three variable three equation case this is Cramer's rule since the $e_1 \wedge e_2 \wedge e_3$ factors of all the wedge products divide out, leaving the familiar determinants.

A numeric example with three equations and two unknowns: In case there are more equations than variables and the equations have a solution, then each of the k-vector quotients will be scalars.

To illustrate here is the solution of a simple example with three equations and two unknowns.

$$\begin{bmatrix}1\\1\\0\end{bmatrix}x+\begin{bmatrix}1\\1\\1\end{bmatrix}y=\begin{bmatrix}1\\1\\2\end{bmatrix}$$

The right wedge product with $(1,1,1)$ solves for x

$$\begin{bmatrix}1\\1\\0\end{bmatrix}\wedge\begin{bmatrix}1\\1\\1\end{bmatrix}x=\begin{bmatrix}1\\1\\2\end{bmatrix}\wedge\begin{bmatrix}1\\1\\1\end{bmatrix}$$

and a left wedge product with $(1,1,0)$ solves for y

$$\begin{bmatrix}1\\1\\0\end{bmatrix}\wedge\begin{bmatrix}1\\1\\1\end{bmatrix}y=\begin{bmatrix}1\\1\\0\end{bmatrix}\wedge\begin{bmatrix}1\\1\\2\end{bmatrix}.$$

Observe that both of these equations have the same factor, so one can compute this only once (if this was zero it would indicate the system of equations has no solution).

Collection of results for x and y yields a Cramer's rule-like form:

$$\begin{bmatrix}x\\y\end{bmatrix}=\frac{1}{(1,1,0)\wedge(1,1,1)}\begin{bmatrix}(1,1,2)\wedge(1,1,1)\\(1,1,0)\wedge(1,1,2)\end{bmatrix}.$$

Writing $e_i \wedge e_j = e_{ij}$, we have the end result:

$$\begin{bmatrix}x\\y\end{bmatrix}=\frac{1}{e_{13}+e_{23}}\begin{bmatrix}-e_{13}-e_{23}\\2e_{13}+2e_{23}\end{bmatrix}=\begin{bmatrix}-1\\2\end{bmatrix}.$$

Equation of a Plane

For the plane of all points $\mathbf{r}$ through the plane passing through three independent points $\mathbf{r}_0$, $\mathbf{r}_1$, and $\mathbf{r}_2$, the normal form of the equation is

$$((\mathbf{r}_2-\mathbf{r}_0)\times(\mathbf{r}_1-\mathbf{r}_0))\cdot(\mathbf{r}-\mathbf{r}_0)=0.$$

The equivalent wedge product equation is

$$(\mathbf{r}_2-\mathbf{r}_0)\wedge(\mathbf{r}_1-\mathbf{r}_0)\wedge(\mathbf{r}-\mathbf{r}_0)=0.$$

Projection and Rejection

Using the Gram–Schmidt process a single vector can be decomposed into two components with respect to a reference vector, namely the projection onto a unit vector in a reference direction, and the difference between the vector and that projection.

With, $\hat{u} = u/\|u\|$, the projection of v onto $\hat{u}$ is

$$Proj_{\hat{u}}\, v = \hat{u}(\hat{u}\cdot v)$$

Orthogonal to that vector is the difference, designated the rejection,

$$v - \hat{u}(\hat{u}\cdot v) = \frac{1}{\|u\|^2}(\|u\|^2\, v - u(u\cdot v))$$

The rejection can be expressed as a single geometric algebraic product in a few different ways

$$\frac{u}{u^2}(uv - u\cdot v) = \frac{1}{u}(u\wedge v) = \hat{u}(\hat{u}\wedge v) = (v\wedge\hat{u})\hat{u}$$

The similarity in form between the projection and the rejection is notable. The sum of these recovers the original vector

$$v = \hat{u}(\hat{u}\cdot v) + \hat{u}(\hat{u}\wedge v)$$

Here the projection is in its customary vector form. An alternate formulation is possible that puts the projection in a form that differs from the usual vector formulation

$$v = \frac{1}{u}(u\cdot v) + \frac{1}{u}(u\wedge v) = (v\cdot u)\frac{1}{u} + (v\wedge u)\frac{1}{u}$$

Working backwards from the end result, it can be observed that this orthogonal decomposition result can in fact follow more directly from the definition of the geometric product itself.

$$v = \hat{u}\hat{u}v = \hat{u}(\hat{u}\cdot v + \hat{u}\wedge v)$$

With this approach, the original geometrical consideration is not necessarily obvious, but it is a much quicker way to get at the same algebraic result.

However, the hint that one can work backwards, coupled with the knowledge that the wedge product can be used to solve sets of linear equations, the problem of orthogonal decomposition can be posed directly,

Let $v = au + x$, where $u\cdot x = 0$. To discard the portions of v that are colinear with u, take the exterior product

$$u\wedge v = u\wedge(au + x) = u\wedge x$$

Here the geometric product can be employed

$$u\wedge v = u\wedge x = ux - u\cdot x = ux$$

Because the geometric product is invertible, this can be solved for x:

$$x = \frac{1}{u}(u\wedge v).$$

The same techniques can be applied to similar problems, such as calculation of the component of a vector in a plane and perpendicular to the plane.

For three dimensions the projective and rejective components of a vector with respect to an arbitrary non-zero unit vector, can be expressed in terms of the dot and cross product

$$\mathbf{v} = (\mathbf{v} \cdot \hat{\mathbf{u}})\hat{\mathbf{u}} + \hat{\mathbf{u}} \times (\mathbf{v} \times \hat{\mathbf{u}}).$$

For the general case the same result can be written in terms of the dot and wedge product and the geometric product of that and the unit vector

$$\mathbf{v} = (\mathbf{v} \cdot \hat{\mathbf{u}})\hat{\mathbf{u}} + (\mathbf{v} \wedge \hat{\mathbf{u}})\hat{\mathbf{u}}.$$

It's also worthwhile to point out that this result can also be expressed using right or left vector division as defined by the geometric product:

$$\mathbf{v} = (\mathbf{v} \cdot \mathbf{u})\frac{1}{\mathbf{u}} + (\mathbf{v} \wedge \mathbf{u})\frac{1}{\mathbf{u}}$$

$$\mathbf{v} = \frac{1}{\mathbf{u}}(\mathbf{u} \cdot \mathbf{v}) + \frac{1}{\mathbf{u}}(\mathbf{u} \wedge \mathbf{v}).$$

Like vector projection and rejection, higher-dimensional analogs of that calculation are also possible using the geometric product.

As an example, one can calculate the component of a vector perpendicular to a plane and the projection of that vector onto the plane.

Let $w = au + bv + x$, where $u \cdot x = v \cdot x = 0$. As above, to discard the portions of w that are colinear with u or v, take the wedge product

$$w \wedge u \wedge v = (au + bv + x) \wedge u \wedge v = x \wedge u \wedge v.$$

Having done this calculation with a vector projection, one can guess that this quantity equals $x(u \wedge v)$. One can also guess there is a vector and bivector dot product like quantity such that the allows the calculation of the component of a vector that is in the "direction of a plane". Both of these guesses are correct, and validating these facts is worthwhile. However, skipping ahead slightly, this to-be-proven fact allows for a nice closed form solution of the vector component outside of the plane:

$$x = (w \wedge u \wedge v)\frac{1}{u \wedge v} = \frac{1}{u \wedge v}(u \wedge v \wedge w).$$

Notice the similarities between this planar rejection result and the vector rejection result. To calculate the component of a vector outside of a plane we take the volume spanned by three vectors (trivector) and "divide out" the plane.

Independent of any use of the geometric product it can be shown that this rejection in terms of the standard basis is

$$x = \frac{1}{(A_{u,v})^2} \sum_{i<j<k} \begin{vmatrix} w_i & w_j & w_k \\ u_i & u_j & u_k \\ v_i & v_j & v_k \end{vmatrix} \begin{vmatrix} u_i & u_j & u_k \\ v_i & v_j & v_k \\ e_i & e_j & e_k \end{vmatrix}$$

where

$$(A_{u,v})^2 = \sum_{i<j} \begin{vmatrix} u_i & u_j \\ v_i & v_j \end{vmatrix} = -(u \wedge v)^2$$

is the squared area of the parallelogram formed by u and v.

The (squared) magnitude of x is

$$\| x \|^2 = x \cdot w = \frac{1}{(A_{u,v})^2} \sum_{i<j<k} \begin{vmatrix} w_i & w_j & w_k \\ u_i & u_j & u_k \\ v_i & v_j & v_k \end{vmatrix}^2$$

Thus, the (squared) volume of the parallelopiped (base area times perpendicular height) is

$$\sum_{i<j<k} \begin{vmatrix} w_i & w_j & w_k \\ u_i & u_j & u_k \\ v_i & v_j & v_k \end{vmatrix}^2$$

Note the similarity in form to the w, u, v trivector itself

$$\sum_{i<j<k} \begin{vmatrix} w_i & w_j & w_k \\ u_i & u_j & u_k \\ v_i & v_j & v_k \end{vmatrix} e_i \wedge e_j \wedge e_k,$$

which, if you take the set of $e_i \wedge e_j \wedge e_k$ as a basis for the trivector space, suggests this is the natural way to define the length of a trivector. Loosely speaking the length of a vector is a length, length of a bivector is area, and the length of a trivector is volume.

If a vector is factored directly into projective and rejective terms using the geometric product $v = \frac{1}{u}(u \cdot v + u \wedge v)$, then it is not necessarily obvious that the rejection term, a product of vector and bivector is even a vector. Expansion of the vector bivector product in terms of the standard basis vectors has the following form

Let $r = \frac{1}{u}(u \wedge v) = \frac{u}{u^2}(u \wedge v) = \frac{1}{\| u \|^2} u(u \wedge v)$

It can be shown that

$$r = \frac{1}{\|u\|^2} \sum_{i<j} \begin{vmatrix} u_i & u_j \\ v_i & v_j \end{vmatrix} \begin{vmatrix} u_i & u_j \\ e_i & e_j \end{vmatrix}$$

(a result that can be shown more easily straight from $r = v - \hat{u}(\hat{u} \cdot v)$).

The rejective term is perpendicular to u, since $\begin{vmatrix} u_i & u_j \\ u_i & u_j \end{vmatrix} = 0$ implies $r \cdot u = 0$.

The magnitude of r is

$$\|r\|^2 = r \cdot v = \frac{1}{\|u\|^2} \sum_{i<j} \begin{vmatrix} u_i & u_j \\ v_i & v_j \end{vmatrix}^2.$$

So, the quantity

$$\|r\|^2 \|u\|^2 = \sum_{i<j} \begin{vmatrix} u_i & u_j \\ v_i & v_j \end{vmatrix}^2$$

is the squared area of the parallelogram formed by u and v.

It is also noteworthy that the bivector can be expressed as

$$u \wedge v = \sum_{i<j} \begin{vmatrix} u_i & u_j \\ v_i & v_j \end{vmatrix} e_i \wedge e_j.$$

Thus is it natural, if one considers each term $e_i \wedge e_j$ as a basis vector of the bivector space, to define the (squared) "length" of that bivector as the (squared) area.

Going back to the geometric product expression for the length of the rejection $\frac{1}{u}(u \wedge v)$ we see that the length of the quotient, a vector, is in this case is the "length" of the bivector divided by the length of the divisor.

This may not be a general result for the length of the product of two ***k***-vectors, however it is a result that may help build some intuition about the significance of the algebraic operations. Namely,

> When a vector is divided out of the plane (parallelogram span) formed from it and another vector, what remains is the perpendicular component of the remaining vector, and its length is the planar area divided by the length of the vector that was divided out.

Area of the Parallelogram Defined by u and v

If A is the area of the parallelogram defined by u and v, then

$$A^2 = \|\mathbf{u} \times \mathbf{v}\|^2 = \sum_{i<j} \begin{vmatrix} u_i & u_j \\ v_i & v_j \end{vmatrix}^2,$$

and

$$A^2 = -(\mathbf{u} \wedge \mathbf{v})^2 = \sum_{i<j} \begin{vmatrix} u_i & u_j \\ v_i & v_j \end{vmatrix}^2.$$

Note that this squared bivector is a geometric multiplication; this computation can alternatively be stated as the Gram determinant of the two vectors.

Angle between two vectors

$$(\sin\theta)^2 = \frac{\|\mathbf{u} \times \mathbf{v}\|^2}{\|\mathbf{u}\|^2 \|\mathbf{v}\|^2}$$

$$(\sin\theta)^2 = -\frac{(\mathbf{u} \wedge \mathbf{v})^2}{\mathbf{u}^2 \mathbf{v}^2}$$

Volume of the Parallelopiped formed by Three Vectors

In vector algebra, the volume of a parallelopiped is given by the square root of the squared norm of the scalar triple product:

$$V^2 = \|(\mathbf{u} \times \mathbf{v}) \cdot \mathbf{w}\|^2 = \begin{vmatrix} u_1 & u_2 & u_3 \\ v_1 & v_2 & v_3 \\ w_1 & w_2 & w_3 \end{vmatrix}^2$$

$$V^2 = -(\mathbf{u} \wedge \mathbf{v} \wedge \mathbf{w})^2 = -\left(\sum_{i<j<k} \begin{vmatrix} u_i & u_j & u_k \\ v_i & v_j & v_k \\ w_i & w_j & w_k \end{vmatrix} \hat{\mathbf{e}}_i \wedge \hat{\mathbf{e}}_j \wedge \hat{\mathbf{e}}_k \right)^2 = \sum_{i<j<k} \begin{vmatrix} u_i & u_j & u_k \\ v_i & v_j & v_k \\ w_i & w_j & w_k \end{vmatrix}^2$$

Product of a Vector and a Bivector

In order to justify the normal to a plane result above, a general examination of the product of a vector and bivector is required. Namely,

$$w(u \wedge v) = \sum_{i,j<k} w_i e_i \begin{vmatrix} u_j & u_k \\ v_j & v_k \end{vmatrix} e_j \wedge e_k$$

This has two parts, the vector part where $i = j$ or $i = k$, and the trivector parts where no indexes equal. After some index summation trickery, and grouping terms and so forth, this is

$$w(u \wedge v) = \sum_{i<j} (w_i e_j - w_j e_i) \begin{vmatrix} u_i & u_j \\ v_i & v_j \end{vmatrix} + \sum_{i<j<k} \begin{vmatrix} w_i & w_j & w_k \\ u_i & u_j & u_k \\ v_i & v_j & v_k \end{vmatrix} e_i \wedge e_j \wedge e_k$$

The trivector term is $w \wedge u \wedge v$. Expansion of $(u \wedge v)w$ yields the same trivector term (it is the completely symmetric part), and the vector term is negated. Like the geometric product of two vectors, this geometric product can be grouped into symmetric and antisymmetric parts, one of which is a pure k-vector. In analogy the antisymmetric part of this product can be called a generalized dot product, and is roughly speaking the dot product of a "plane" (bivector), and a vector.

The properties of this generalized dot product remain to be explored, but first here is a summary of the notation

$$w(u \wedge v) = w \cdot (u \wedge v) + w \wedge u \wedge v$$

$$(u \wedge v)w = -w \cdot (u \wedge v) + w \wedge u \wedge v$$

$$w \wedge u \wedge v = \frac{1}{2}(w(u \wedge v) + (u \wedge v)w)$$

$$w \cdot (u \wedge v) = \frac{1}{2}(w(u \wedge v) - (u \wedge v)w)$$

Let $w = x + y$, where $x = au + bv$, and $y \cdot u = y \cdot v = 0$. Expressing w and the $u \wedge v$, products in terms of these components is

$$w(u \wedge v) = x(u \wedge v) + y(u \wedge v) = x \cdot (u \wedge v) + y \cdot (u \wedge v) + y \wedge u \wedge v$$

With the conditions and definitions above, and some manipulation, it can be shown that the term $y \cdot (u \wedge v) = 0$, which then justifies the previous solution of the normal to a plane problem. Since the vector term of the vector bivector product the name dot product is zero when the vector is perpendicular to the plane (bivector), and this vector, bivector "dot product" selects only the components that are in the plane, so in analogy to the vector-vector dot product this name itself is justified by more than the fact this is the non-wedge product term of the geometric vector-bivector product.

Derivative of a Unit Vector

It can be shown that a unit vector derivative can be expressed using the cross product

$$\frac{d}{dt}\left(\frac{\mathbf{r}}{\|\mathbf{r}\|}\right) = \frac{1}{\|\mathbf{r}\|^3}\left(\mathbf{r} \times \frac{d\mathbf{r}}{dt}\right) \times \mathbf{r} = \left(\hat{\mathbf{r}} \times \frac{1}{\|\mathbf{r}\|}\frac{d\mathbf{r}}{dt}\right) \times \hat{\mathbf{r}}$$

The equivalent geometric product generalization is

$$\frac{d}{dt}\left(\frac{\mathbf{r}}{\|\mathbf{r}\|}\right) = \frac{1}{\|\mathbf{r}\|^3}\mathbf{r}\left(\mathbf{r} \wedge \frac{d\mathbf{r}}{dt}\right) = \frac{1}{\mathbf{r}}\left(\hat{\mathbf{r}} \wedge \frac{d\mathbf{r}}{dt}\right)$$

Thus this derivative is the component of $\frac{1}{\|\mathbf{r}\|}\frac{d\mathbf{r}}{dt}$ in the direction perpendicular to $\mathbf{r}$. In other words, this is $\frac{1}{\|\mathbf{r}\|}\frac{d\mathbf{r}}{dt}$ minus the projection of that vector onto $\hat{\mathbf{r}}$.

This intuitively makes sense (but a picture would help) since a unit vector is constrained to circular motion, and any change to a unit vector due to a change in its generating vector has to be in the direction of the rejection of $\hat{\mathbf{r}}$ from $\frac{d\mathbf{r}}{dt}$. That rejection has to be scaled by 1/|r| to get the final result.

When the objective isn't comparing to the cross product, it's also notable that this unit vector derivative can be written

$$\mathbf{r}\frac{d\hat{\mathbf{r}}}{dt} = \hat{\mathbf{r}} \wedge \frac{d\mathbf{r}}{dt}$$

References

- Ricci-Curbastro, Gregorio; Levi-Civita, Tullio (1900), "Méthodes de calcul différentiel absolu et leurs applications", Mathematische Annalen, 54 (1): 125–201, doi:10.1007/BF01454201, ISSN 1432-1807
- Klawitter, Daniel (2014), Clifford Algebras: Geometric Modelling and Chain Geometries with Application in Kinematics, Springer, ISBN 9783658076184
- Lasenby, Joan; Lasenby, Anthony N.; Doran, Chris J. L. (2000), "A Unified Mathematical Language for Physics and Engineering in the 21st Century" (PDF), Philosophical Transactions of the Royal Society of London, pp. 1–18, 358 (A 358): 21, Bibcode:2000RSPTA.358...21L, doi:10.1098/rsta.2000.0517
- Hestenes, David (2003), "Oersted Medal Lecture 2002: Reforming the Mathematical Language of Physics" (PDF), Am. J. Phys., 71 (2): 104–121, Bibcode:2003AmJPh..71..104H, doi:10.1119/1.1522700

2

Abstract Algebra

The study of algebraic structures, such as groups, rings, fields, lattices, modules, etc. is under the domain of abstract algebra. The concepts of sets, binary operations, inverse elements and identity elements are fundamental concepts in this field. This chapter has been carefully written to provide an easy understanding of the varied algebraic structures studied under abstract algebra.

Abstract Algebra

Abstract algebra is a part of math which studies algebraic structures. These include:

- groups
- rings
- fields
- vector spaces
- algebras
- Boolean algebra

It is normal to build a theory on one kind of structure, like group theory or category theory.

The purpose of the theory of each concept is to organize the precise definition of the concept, examples of it, its substructures, the ways to relate different examples of the concept algebraically (these are called morphisms in some cases), and the concept's applications, both inside its own theory and outside in other areas of mathematics.

During history, different fields of mathematics have used algebras. Algebras are about finding or specifying rules on how to calculate with certain mathematical formulas and expressions. Another algebra (which is not *abstract*) is elementary algebra, for example.

Examples

- Solving equations with many variables. This leads to matrices, determinants and linear algebra.
- Finding formulas for polynomial equations. This led to the discovery of groups, as an expression of symmetry.
- Quadratic and higher-degree equations and Diophantine equations - especially when Fermat's last theorem was proved led to the definition of rings, and ideals.

Algebraic Structure

In mathematics, and more specifically in abstract algebra, an algebraic structure on a set A (called carrier set or underlying set) is a collection of finitary operations on A; the set A with this structure is also called an algebra.

Examples of algebraic structures include groups, rings, fields, and lattices. More complex structures can be defined by introducing multiple operations, different underlying sets, or by altering the defining axioms. Examples of more complex algebraic structures include vector spaces, modules, and algebras.

The properties of specific algebraic structures are studied in abstract algebra. The general theory of algebraic structures has been formalized in universal algebra. The language of category theory is used to express and study relationships between different classes of algebraic and non-algebraic objects. This is because it is sometimes possible to find strong connections between some classes of objects, sometimes of different kinds. For example, Galois theory establishes a connection between certain fields and groups: two algebraic structures of different kinds.

Introduction

Addition and multiplication on numbers are the prototypical example of an operation that combines two elements of a set to produce a third. These operations obey several algebraic laws. For example, $a + (b + c) = (a + b) + c$ and $a(bc) = (ab)c$, both examples of the *associative law*. Also $a + b = b + a$, and $ab = ba$, the *commutative law*. Many systems studied by mathematicians have operations that obey some, but not necessarily all, of the laws of ordinary arithmetic. For example, rotations of objects in three-dimensional space can be combined by performing the first rotation and then applying the second rotation to the object in its new orientation. This operation on rotations obeys the associative law, but can fail the commutative law.

Mathematicians give names to sets with one or more operations that obey a particular collection of laws, and study them in the abstract as algebraic structures. When a new problem can be shown to follow the laws of one of these algebraic structures, all the work that has been done on that category in the past can be applied to the new problem.

In full generality, algebraic structures may involve an arbitrary number of sets and operations that can combine more than two elements (higher arity), but this article focuses on binary operations on one or two sets. The examples here are by no means a complete list, but they are meant to be a representative list and include the most common structures. Longer lists of algebraic structures may be found in the external links and within *Category:Algebraic structures*. Structures are listed in approximate order of increasing complexity.

Examples

One Set with Operations

Simple structures: no binary operation:

- Set: a degenerate algebraic structure S having no operations.

- Pointed set: S has one or more distinguished elements, often 0, 1, or both.
- Unary system: S and a single unary operation over S.
- Pointed unary system: a unary system with S a pointed set.

Group-like structures: one binary operation. The binary operation can be indicated by any symbol, or with no symbol (juxtaposition) as is done for ordinary multiplication of real numbers.

- Magma or groupoid: S and a single binary operation over S.
- Semigroup: an associative magma.
- Monoid: a semigroup with identity.
- Group: a monoid with a unary operation (inverse), giving rise to inverse elements.
- Abelian group: a group whose binary operation is commutative.
- Semilattice: a semigroup whose operation is idempotent and commutative. The binary operation can be called either meet or join.
- Quasigroup: a magma obeying the latin square property. A quasigroup may also be represented using three binary operations.
- Loop: a quasigroup with identity.

Ring-like structures or Ringoids: two binary operations, often called addition and multiplication, with multiplication distributing over addition.

- Semiring: a ringoid such that S is a monoid under each operation. Addition is typically assumed to be commutative and associative, and the monoid product is assumed to distribute over the addition on both sides, and the additive identity satisfies $0\,x = 0$ for all x.
- Near-ring: a semiring whose additive monoid is a (not necessarily abelian) group.
- Ring: a semiring whose additive monoid is an abelian group.
- Lie ring: a ringoid whose additive monoid is an abelian group, but whose multiplicative operation satisfies the Jacobi identity rather than associativity.
- Boolean ring: a commutative ring with idempotent multiplication operation.
- Field: a commutative ring which contains a multiplicative inverse for every nonzero element
- Kleene algebras: a semiring with idempotent addition and a unary operation, the Kleene star, satisfying additional properties.
- *-algebra: a ring with an additional unary operation (*) satisfying additional properties.

Lattice structures: two or more binary operations, including operations called meet and join, connected by the absorption law.

- Complete lattice: a lattice in which arbitrary meet and joins exist.

- Bounded lattice: a lattice with a greatest element and least element.
- Complemented lattice: a bounded lattice with a unary operation, complementation, denoted by postfix $^{\perp}$. The join of an element with its complement is the greatest element, and the meet of the two elements is the least element.
- Modular lattice: a lattice whose elements satisfy the additional *modular identity*.
- Distributive lattice: a lattice in which each of meet and join distributes over the other. Distributive lattices are modular, but the converse does not hold.
- Boolean algebra: a complemented distributive lattice. Either of meet or join can be defined in terms of the other and complementation. This can be shown to be equivalent with the ring-like structure of the same name above.
- Heyting algebra: a bounded distributive lattice with an added binary operation, relative pseudo-complement, denoted by infix →, and governed by the axioms $x \to x = 1$, $x(x \to y) = xy$, $y(x \to y) = y$, $x \to (yz) = (x \to y)(x \to z)$.

Arithmetics: two binary operations, addition and multiplication. S is an infinite set. Arithmetics are pointed unary systems, whose unary operation is injective successor, and with distinguished element 0.

- Robinson arithmetic. Addition and multiplication are recursively defined by means of successor. 0 is the identity element for addition, and annihilates multiplication. Robinson arithmetic is listed here even though it is a variety, because of its closeness to Peano arithmetic.
- Peano arithmetic. Robinson arithmetic with an axiom schema of induction. Most ring and field axioms bearing on the properties of addition and multiplication are theorems of Peano arithmetic or of proper extensions thereof.

Two Sets with Operations

Module-like structures: composite systems involving two sets and employing at least two binary operations.

- Group with operators: a group G with a set Ω and a binary operation $\Omega \times G \to G$ satisfying certain axioms.
- Module: an abelian group M and a ring R acting as operators on M. The members of R are sometimes called scalars, and the binary operation of *scalar multiplication* is a function $R \times M \to M$, which satisfies several axioms. Counting the ring operations these systems have at least three operations.
- Vector space: a module where the ring R is a division ring or field.
- Graded vector space: a vector space with a direct sum decomposition breaking the space into "grades".
- Quadratic space: a vector space V over a field F with a function from V into F satisfying certain properties. Every quadratic space is also an inner product space.

Algebra-like structures: composite system defined over two sets, a ring R and a R module M equipped with an operation called multiplication. This can be viewed as a system with five binary operations: two operations on R, two on M and one involving both R and M.

- Algebra over a ring (also *R-algebra*): a module over a commutative ring R, which also carries a multiplication operation that is compatible with the module structure. This includes distributivity over addition and linearity with respect to multiplication by elements of R. The theory of an algebra over a field is especially well developed.
- Associative algebra: an algebra over a ring such that the multiplication is associative.
- Nonassociative algebra: a module over a commutative ring, equipped with a ring multiplication operation that is not necessarily associative. Often associativity is replaced with a different identity, such as alternation, the Jacobi identity, or the Jordan identity.
- Coalgebra: a vector space with a "comultiplication" defined dually to that of associative algebras.
- Lie algebra: a special type of nonassociative algebra whose product satisfies the Jacobi identity.
- Lie coalgebra: a vector space with a "comultiplication" defined dually to that of Lie algebras.
- Graded algebra: a graded vector space with an algebra structure compatible with the grading. The idea is that if the grades of two elements a and b are known, then the grade of ab is known, and so the location of the product ab is determined in the decomposition.
- Inner product space: an F vector space V with a sesquilinear binary operation $V \times V \rightarrow F$.

Four or more binary operations:

- Bialgebra: an associative algebra with a compatible coalgebra structure.
- Lie bialgebra: a Lie algebra with a compatible bialgebra structure.
- Hopf algebra: a bialgebra with a connection axiom (antipode).
- Clifford algebra: a graded associative algebra equipped with an exterior product from which may be derived several possible inner products. Exterior algebras and geometric algebras are special cases of this construction.

Hybrid Structures

Algebraic structures can also coexist with added structure of non-algebraic nature, such as partial order or a topology. The added structure must be compatible, in some sense, with the algebraic structure.

- Topological group: a group with a topology compatible with the group operation.
- Lie group: a topological group with a compatible smooth manifold structure.
- Ordered groups, ordered rings and ordered fields: each type of structure with a compatible partial order.

- Archimedean group: a linearly ordered group for which the Archimedean property holds.
- Topological vector space: a vector space whose *M* has a compatible topology.
- Normed vector space: a vector space with a compatible norm. If such a space is complete (as a metric space) then it is called a Banach space.
- Hilbert space: an inner product space over the real or complex numbers whose inner product gives rise to a Banach space structure.
- Vertex operator algebra
- Von Neumann algebra: a *-algebra of operators on a Hilbert space equipped with the weak operator topology.

Universal Algebra

Algebraic structures are defined through different configurations of axioms. Universal algebra abstractly studies such objects. One major dichotomy is between structures that are axiomatized entirely by *identities* and structures that are not. If all axioms defining a class of algebras are identities, then the class of objects is a variety.

Identities are equations formulated using only the operations the structure allows, and variables that are tacitly universally quantified over the relevant universe. Identities contain no connectives, existentially quantified variables, or relations of any kind other than the allowed operations. The study of varieties is an important part of universal algebra. An algebraic structure in a variety may be understood as the quotient algebra of term algebra (also called "absolutely free algebra") divided by the equivalence relations generated by a set of identities. So, a collection of functions with given signatures generate a free algebra, the term algebra T. Given a set of equational identities (the axioms), one may consider their symmetric, transitive closure E. The quotient algebra T/E is then the algebraic structure or variety. Thus, for example, groups have a signature containing two operators: the multiplication operator m, taking two arguments, and the inverse operator i, taking one argument, and the identity element e, a constant, which may be considered an operator that takes zero arguments. Given a (countable) set of variables x, y, z, etc. the term algebra is the collection of all possible terms involving m, i, e and the variables; so for example, $m(i(x), m(x,m(y,e)))$ would be an element of the term algebra. One of the axioms defining a group is the identity $m(x, i(x)) = e$; another is $m(x,e) = x$. The axioms can be represented as trees. These equations induce equivalence classes on the free algebra; the quotient algebra then has the algebraic structure of a group.

Several non-variety structures fail to be varieties, because either:

1. It is necessary that $0 \neq 1$, 0 being the additive identity element and 1 being a multiplicative identity element, but this is a nonidentity;
2. Structures such as fields have some axioms that hold only for nonzero members of S. For an algebraic structure to be a variety, its operations must be defined for *all* members of S; there can be no partial operations.

Structures whose axioms unavoidably include nonidentities are among the most important ones in mathematics, e.g., fields and division rings. Although structures with nonidentities retain an undoubted algebraic flavor, they suffer from defects varieties do not have. For example, the product of two fields is not a field.

Category Theory

Category theory is another tool for studying algebraic structures. A category is a collection of *objects* with associated *morphisms*. Every algebraic structure has its own notion of homomorphism, namely any function compatible with the operation(s) defining the structure. In this way, every algebraic structure gives rise to a category. For example, the category of groups has all groups as objects and all group homomorphisms as morphisms. This concrete category may be seen as a category of sets with added category-theoretic structure. Likewise, the category of topological groups (whose morphisms are the continuous group homomorphisms) is a category of topological spaces with extra structure. A forgetful functor between categories of algebraic structures "forgets" a part of a structure.

There are various concepts in category theory that try to capture the algebraic character of a context, for instance

- algebraic category
- essentially algebraic category
- presentable category
- locally presentable category
- monadic functors and categories
- universal property.

Different Meanings of "Structure"

In a slight abuse of notation, the word "structure" can also refer to just the operations on a structure, instead of the underlying set itself. For example, the sentence, "We have defined a ring *structure* on the set A," means that we have defined ring *operations* on the set A. For another example, the group $(\mathbb{Z},+)$ can be seen as a set $\mathbb{Z}$ that is equipped with an *algebraic structure,* namely the *operation* $+$.

Group

In mathematics, a group is an algebraic structure consisting of a set of elements equipped with an operation that combines any two elements to form a third element and that satisfies four conditions called the group axioms, namely closure, associativity, identity and invertibility. One of the most familiar examples of a group is the set of integers together with the addition operation, but the abstract formalization of the group axioms, detached as it is from the concrete nature of any particular group and its operation, applies much more widely. It allows entities with highly diverse mathematical origins in abstract algebra and beyond to be handled in a flexible way while retaining their essential structural aspects. The ubiquity of groups in numerous areas within and outside mathematics makes them a central organizing principle of contemporary mathematics.

Groups share a fundamental kinship with the notion of symmetry. For example, a symmetry group encodes symmetry features of a geometrical object: the group consists of the set of transformations that leave the object unchanged and the operation of combining two such transformations by performing one after the other. Lie groups are the symmetry groups used in the Standard Model of particle physics; Poincaré groups, which are also Lie groups, can express the physical symmetry underlying special relativity; and point groups are used to help understand symmetry phenomena in molecular chemistry.

The concept of a group arose from the study of polynomial equations, starting with Évariste Galois in the 1830s. After contributions from other fields such as number theory and geometry, the group notion was generalized and firmly established around 1870. Modern group theory—an active mathematical discipline—studies groups in their own right. To explore groups, mathematicians have devised various notions to break groups into smaller, better-understandable pieces, such as subgroups, quotient groups and simple groups. In addition to their abstract properties, group theorists also study the different ways in which a group can be expressed concretely, both from a point of view of representation theory (that is, through the representations of the group) and of computational group theory. A theory has been developed for finite groups, which culminated with the classification of finite simple groups, completed in 2004. Since the mid-1980s, geometric group theory, which studies finitely generated groups as geometric objects, has become a particularly active area in group theory.

Definition and Illustration

First Example: the Integers

One of the most familiar groups is the set of integers Z which consists of the numbers

$$\ldots, -4, -3, -2, -1, 0, 1, 2, 3, 4, \ldots, \text{ together with addition.}$$

The following properties of integer addition serve as a model for the abstract group axioms given in the definition below.

- For any two integers a and b, the sum $a + b$ is also an integer. That is, addition of integers always yields an integer. This property is known as *closure* under addition.
- For all integers a, b and c, $(a + b) + c = a + (b + c)$. Expressed in words, adding a to b first, and then adding the result to c gives the same final result as adding a to the sum of b and c, a property known as *associativity*.
- If a is any integer, then $0 + a = a + 0 = a$. Zero is called the *identity element* of addition because adding it to any integer returns the same integer.
- For every integer a, there is an integer b such that $a + b = b + a = 0$. The integer b is called the *inverse element* of the integer a and is denoted $-a$.

The integers, together with the operation +, form a mathematical object belonging to a broad class sharing similar structural aspects. To appropriately understand these structures as a collective, the following abstract definition is developed.

Definition

"The axioms for a group are short and natural... Yet somehow hidden behind these axioms is the monster simple group, a huge and extraordinary mathematical object, which appears to rely on numerous bizarre coincidences to exist. The axioms for groups give no obvious hint that anything like this exists."

Richard Borcherds in Mathematicians: An Outer View of the Inner World

A group is a set, G, together with an operation • (called the *group law* of G) that combines any two elements a and b to form another element, denoted $a \bullet b$ or ab. To qualify as a group, the set and operation, $(G, \bullet)$, must satisfy four requirements known as the *group axioms*:

Closure

For all a, b in G, the result of the operation, $a \bullet b$, is also in G.

Associativity

For all a, b and c in G, $(a \bullet b) \bullet c = a \bullet (b \bullet c)$.

Identity element

There exists an element e in G such that, for every element a in G, the equation $e \bullet a = a \bullet e = a$ holds. Such an element is unique, and thus one speaks of *the* identity element.

Inverse element

For each a in G, there exists an element b in G, commonly denoted a^{-1} (or $-a$, if the operation is denoted "+"), such that $a \bullet b = b \bullet a = e$, where e is the identity element.

The result of an operation may depend on the order of the operands. In other words, the result of combining element a with element b need not yield the same result as combining element b with element a; the equation

$$a \bullet b = b \bullet a$$

may not always be true. This equation always holds in the group of integers under addition, because $a + b = b + a$ for any two integers (commutativity of addition). Groups for which the commutativity equation $a \bullet b = b \bullet a$ always holds are called *abelian groups* (in honor of Niels Henrik Abel). The symmetry group described in the following section is an example of a group that is not abelian.

The identity element of a group G is often written as 1 or 1_G, a notation inherited from the multiplicative identity. If a group is abelian, then one may choose to denote the group operation by + and the identity element by 0; in that case, the group is called an additive group. The identity element can also be written as *id*.

The set G is called the *underlying set* of the group $(G, \bullet)$. Often the group's underlying set G is used as a short name for the group $(G, \bullet)$. Along the same lines, shorthand expressions such as "a subset of the group G" or "an element of group G" are used when what is actually meant is "a subset of the underlying set G of the group $(G, \bullet)$" or "an element of the underlying set G of the group $(G, \bullet)$". Usually, it is clear from the context whether a symbol like G refers to a group or to an underlying set.

Second Example: a Symmetry Group

Two figures in the plane are congruent if one can be changed into the other using a combination of rotations, reflections, and translations. Any figure is congruent to itself. However, some figures are congruent to themselves in more than one way, and these extra congruences are called symmetries. A square has eight symmetries. These are:

The elements of the symmetry group of the square (D4). Vertices are identified by color or number.

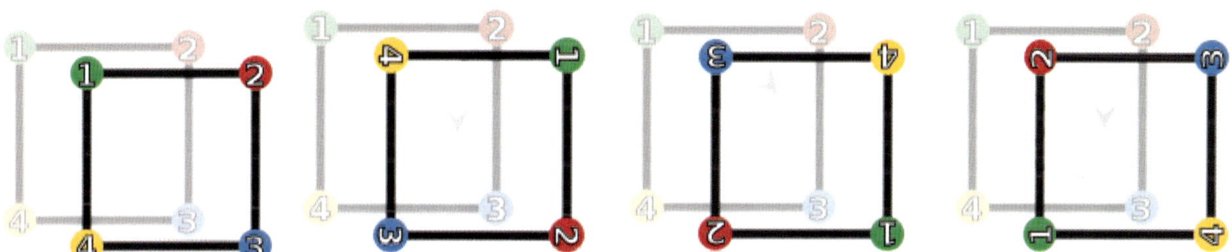

id (keeping it as it is) | r_1 (rotation by 90° clockwise) | r_2 (rotation by 180° clockwise) | r_3 (rotation by 270° clockwise)

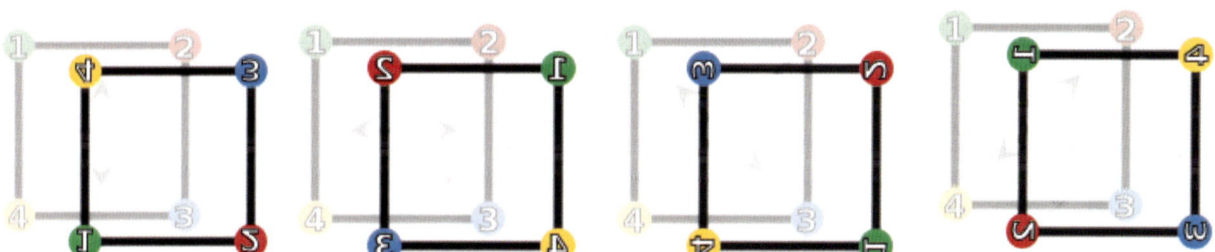

f_v (vertical reflection) | f_h (horizontal reflection) | f_d (diagonal reflection) | f_c (counter-diagonal reflection)

- the identity operation leaving everything unchanged, denoted id;
- rotations of the square around its center by 90° clockwise, 180° clockwise, and 270° clockwise, denoted by r_1, r_2 and r_3, respectively;
- reflections about the vertical and horizontal middle line (f_h and f_v), or through the two diagonals (f_d and f_c).

These symmetries are represented by functions. Each of these functions sends a point in the square to the corresponding point under the symmetry. For example, r_1 sends a point to its rotation 90° clockwise around the square's center, and f_h sends a point to its reflection across the square's vertical middle line. Composing two of these symmetry functions gives another symmetry function. These symmetries determine a group called the dihedral group of degree 4 and denoted D_4. The underlying set of the group is the above set of symmetry functions, and the group operation is function composition. Two symmetries are combined by composing them as functions, that is, applying the first one to the square, and the second one to the result of the first application. The result of performing first a and then b is written symbolically *from right to left* as

$b \cdot a$ ("apply the symmetry b after performing the symmetry a").

The right-to-left notation is the same notation that is used for composition of functions.

The group table on the right lists the results of all such compositions possible. For example, rotating by 270° clockwise (r_3) and then reflecting horizontally (f_h) is the same as performing a reflection along the diagonal (f_d). Using the above symbols, highlighted in blue in the group table:

$f_h \cdot r_3 = f_d$.

Group table of D_4

•	**id**	$\mathbf{r_1}$	$\mathbf{r_2}$	$\mathbf{r_3}$	$\mathbf{f_v}$	$\mathbf{f_h}$	$\mathbf{f_d}$	$\mathbf{f_c}$
id	id	r_1	r_2	r_3	f_v	f_h	f_d	f_c
$\mathbf{r_1}$	r_1	r_2	r_3	id	f_c	f_d	f_v	f_h
$\mathbf{r_2}$	r_2	r_3	id	r_1	f_h	f_v	f_c	f_d
$\mathbf{r_3}$	r_3	id	r_1	r_2	f_d	f_c	f_h	[redacted]
$\mathbf{f_v}$	f_v	f_d	f_h	f_c	id	r_2	r_1	r_3
$\mathbf{f_h}$	f_h	f_c	f_v	f_d	r_2	id	r_3	r_1
$\mathbf{f_d}$	f_d	f_h	f_c	f_v	r_3	r_1	id	r_2
$\mathbf{f_c}$	f_c	f_v	f_d	f_h	r_1	r_3	r_2	id

The elements id, r_1, r_2, and r_3 form a subgroup, highlighted in red (upper left region). A left and right coset of this subgroup is highlighted in green (in the last row) and yellow (last column), respectively.

Given this set of symmetries and the described operation, the group axioms can be understood as follows:

1. The closure axiom demands that the composition $b \cdot a$ of any two symmetries a and b is also a symmetry. Another example for the group operation is

   $r_3 \cdot f_h = f_c$,

   i.e., rotating 270° clockwise after reflecting horizontally equals reflecting along the counter-diagonal (f_c). Indeed every other combination of two symmetries still gives a symmetry, as can be checked using the group table.

2. The associativity constraint deals with composing more than two symmetries: Starting with three elements a, b and c of D_4, there are two possible ways of using these three symmetries in this order to determine a symmetry of the square. One of these ways is to first compose a and b into a single symmetry, then to compose that symmetry with c. The other way is to first compose b and c, then to compose the resulting symmetry with a. The associativity condition

   $(a \cdot b) \cdot c = a \cdot (b \cdot c)$

   means that these two ways are the same, i.e., a product of many group elements can be simplified in any grouping. For example, $(f_d \cdot f_v) \cdot r_2 = f_d \cdot (f_v \cdot r_2)$ can be checked using the group table at the right

$$(f_d \bullet f_v) \bullet r_2 \quad = \quad r_3 \bullet r_2 \quad = \quad r_1, \text{ which equals}$$

$$f_d \bullet (f_v \bullet r_2) \quad = \quad f_d \bullet f_h \quad = \quad r_1.$$

While associativity is true for the symmetries of the square and addition of numbers, it is not true for all operations. For instance, subtraction of numbers is not associative: (7 − 3) − 2 = 2 is not the same as 7 − (3 − 2) = 6.

3. The identity element is the symmetry id leaving everything unchanged: for any symmetry *a*, performing id after *a* (or *a* after id) equals *a*, in symbolic form,

   $\mathrm{id} \bullet a = a,$

   $a \bullet \mathrm{id} = a.$

4. An inverse element undoes the transformation of some other element. Every symmetry can be undone: each of the following transformations—identity id, the reflections f_h, f_v, f_d, f_c and the 180° rotation r_2—is its own inverse, because performing it twice brings the square back to its original orientation. The rotations r_3 and r_1 are each other's inverses, because rotating 90° and then rotation 270° (or vice versa) yields a rotation over 360° which leaves the square unchanged. In symbols,

   $f_h \bullet f_h = \mathrm{id},$

   $r_3 \bullet r_1 = r_1 \bullet r_3 = \mathrm{id}.$

In contrast to the group of integers above, where the order of the operation is irrelevant, it does matter in D_4: $f_h \bullet r_1 = f_c$ but $r_1 \bullet f_h = f_d$. In other words, D_4 is not abelian, which makes the group structure more difficult than the integers introduced first.

Elementary Consequences of the Group Axioms

Basic facts about all groups that can be obtained directly from the group axioms are commonly subsumed under *elementary group theory*. For example, repeated applications of the associativity axiom show that the unambiguity of

$$a \bullet b \bullet c = (a \bullet b) \bullet c = a \bullet (b \bullet c)$$

generalizes to more than three factors. Because this implies that parentheses can be inserted anywhere within such a series of terms, parentheses are usually omitted.

The axioms may be weakened to assert only the existence of a left identity and left inverses. Both can be shown to be actually two-sided, so the resulting definition is equivalent to the one given above.

Uniqueness of Identity Element and Inverses

Two important consequences of the group axioms are the uniqueness of the identity element and the uniqueness of inverse elements. There can be only one identity element in a group, and each element in a group has exactly one inverse element. Thus, it is customary to speak of *the* identity, and *the* inverse of an element.

To prove the uniqueness of an inverse element of a, suppose that a has two inverses, denoted b and c, in a group $(G, \bullet)$. Then

b	$=$	$b \bullet e$	as e is the identity element
	$=$	$b \bullet (a \bullet c)$	because c is an inverse of a, so $e = a \bullet c$
	$=$	$(b \bullet a) \bullet c$	by associativity, which allows rearranging the parentheses
	$=$	$e \bullet c$	since b is an inverse of a, i.e., $b \bullet a = e$
	$=$	c	for e is the identity element

The term b on the first line above and the c on the last are equal, since they are connected by a chain of equalities. In other words, there is only one inverse element of a. Similarly, to prove that the identity element of a group is unique, assume G is a group with two identity elements e and f. Then $e = e \bullet f = f$, hence e and f are equal.

Division

In groups, the existence of inverse elements implies that division is possible: given elements a and b of the group G, there is exactly one solution x in G to the equation $x \bullet a = b$, namely $b \bullet a^{-1}$. In fact, we have

$$(b \bullet a^{-1}) \bullet a = b \bullet (a^{-1} \bullet a) = b \bullet e = b.$$

Uniqueness results by multiplying the two sides of the equation $x \bullet a = b$ by a^{-1}. The element $b \bullet a^{-1}$, often denoted b / a, is called the *right quotient* of b by a, or the result of the *right division* of b by a.

Similarly there is exactly one solution y in G to the equation $a \bullet y = b$, namely $y = a^{-1} \bullet b$. This solution is the *left quotient* of b by a, and is sometimes denoted $a \backslash b$.

In general b / a and $a \backslash b$ may be different, but, if the group operation is commutative (that is, if the group is abelian), they are equal. In this case, the group operation is often denoted as an addition, and one talks of *subtraction* and *difference* instead of division and quotient.

A consequence of this is that multiplication by a group element g is a bijection. Specifically, if g is an element of the group G, the function (mathematics) from G to itself that maps $h \in G$ to $g \bullet h$ is a bijection. This function is called the *left translation* by g . Similarly, the *right translation* by g is the bijection from G to itself, that maps h to $h \bullet g$. If G is abelian, the left and the right translation by a group element are the same.

Basic Concepts

The following sections use mathematical symbols such as $X = \{x, y, z\}$ to denote a set X containing elements x, y, and z, or alternatively $x \in X$ to restate that x is an element of X. The notation $f : X \to Y$ means f is a function assigning to every element of X an element of Y.

To understand groups beyond the level of mere symbolic manipulations as above, more structural concepts have to be employed. There is a conceptual principle underlying all of the following notions: to take advantage of the structure offered by groups (which sets, being "structureless", do not have), constructions related to groups have to be *compatible* with the group operation. This compatibility manifests itself in the following notions in various ways. For example, groups can be related to each other via functions called group homomorphisms. By the mentioned principle, they are required to respect the group structures in a precise sense. The structure of groups can also be understood by breaking them into pieces called subgroups and quotient groups. The principle of "preserving structures"—a recurring topic in mathematics throughout—is an instance of working in a category, in this case the category of groups.

Group Homomorphisms

Group homomorphisms are functions that preserve group structure. A function $a: G \to H$ between two groups $(G, \bullet)$ and $(H, *)$ is called a *homomorphism* if the equation

$$a(g \bullet k) = a(g) * a(k)$$

holds for all elements g, k in G. In other words, the result is the same when performing the group operation after or before applying the map a. This requirement ensures that $a(1_G) = 1_H$, and also $a(g)^{-1} = a(g^{-1})$ for all g in G. Thus a group homomorphism respects all the structure of G provided by the group axioms.

Two groups G and H are called *isomorphic* if there exist group homomorphisms $a: G \to H$ and $b: H \to G$, such that applying the two functions one after another in each of the two possible orders gives the identity functions of G and H. That is, $a(b(h)) = h$ and $b(a(g)) = g$ for any g in G and h in H. From an abstract point of view, isomorphic groups carry the same information. For example, proving that $g \bullet g = 1_G$ for some element g of G is equivalent to proving that $a(g) * a(g) = 1_H$, because applying a to the first equality yields the second, and applying b to the second gives back the first.

Subgroups

Informally, a *subgroup* is a group H contained within a bigger one, G. Concretely, the identity element of G is contained in H, and whenever h_1 and h_2 are in H, then so are $h_1 \bullet h_2$ and h_1^{-1}, so the elements of H, equipped with the group operation on G restricted to H, indeed form a group.

In the example above, the identity and the rotations constitute a subgroup $R = \{\mathrm{id}, r_1, r_2, r_3\}$, highlighted in red in the group table above: any two rotations composed are still a rotation, and a rotation can be undone by (i.e., is inverse to) the complementary rotations 270° for 90°, 180° for 180°, and 90° for 270° (note that rotation in the opposite direction is not defined). The subgroup test is a necessary and sufficient condition for a nonempty subset H of a group G to be a subgroup: it is sufficient to check that $g^{-1}h \in H$ for all elements $g, h \in H$. Knowing the subgroups is important in understanding the group as a whole.

Given any subset S of a group G, the subgroup generated by S consists of products of elements of S and their inverses. It is the smallest subgroup of G containing S. In the introductory example above, the subgroup generated by r_2 and f_v consists of these two elements, the identity element id and $f_h = f_v \bullet r_2$. Again, this is a subgroup, because combining any two of these four elements or their inverses (which are, in this particular case, these same elements) yields an element of this subgroup.

Cosets

In many situations it is desirable to consider two group elements the same if they differ by an element of a given subgroup. For example, in D_4 above, once a reflection is performed, the square never gets back to the r_2 configuration by just applying the rotation operations (and no further reflections), i.e., the rotation operations are irrelevant to the question whether a reflection has been performed. Cosets are used to formalize this insight: a subgroup H defines left and right cosets, which can be thought of as translations of H by arbitrary group elements g. In symbolic terms, the *left* and *right* cosets of H containing g are

$$gH = \{g \cdot h : h \in H\} \text{ and } Hg = \{h \cdot g : h \in H\}, \text{ respectively.}$$

The left cosets of any subgroup H form a partition of G; that is, the union of all left cosets is equal to G and two left cosets are either equal or have an empty intersection. The first case $g_1H = g_2H$ happens precisely when $g_1^{-1} \cdot g_2 \in H$, i.e., if the two elements differ by an element of H. Similar considerations apply to the right cosets of H. The left and right cosets of H may or may not be equal. If they are, i.e., for all g in G, $gH = Hg$, then H is said to be a *normal subgroup*.

In D_4, the introductory symmetry group, the left cosets gR of the subgroup R consisting of the rotations are either equal to R, if g is an element of R itself, or otherwise equal to $U = f_cR = \{f_c, f_v, f_d, f_h\}$ (highlighted in green). The subgroup R is also normal, because $f_cR = U = Rf_c$ and similarly for any element other than f_c. (In fact, in the case of D_4, observe that all such cosets are equal, such that $f_hR = f_vR = f_dR = f_cR$.)

Quotient Groups

In some situations the set of cosets of a subgroup can be endowed with a group law, giving a *quotient group* or *factor group*. For this to be possible, the subgroup has to be normal. Given any normal subgroup N, the quotient group is defined by

$$G / N = \{gN, g \in G\}, \text{“}G \text{ modulo } N\text{”}.$$

This set inherits a group operation (sometimes called coset multiplication, or coset addition) from the original group G: $(gN) \cdot (hN) = (gh)N$ for all g and h in G. This definition is motivated by the idea (itself an instance of general structural considerations outlined above) that the map $G \to G / N$ that associates to any element g its coset gN be a group homomorphism, or by general abstract considerations called universal properties. The coset $eN = N$ serves as the identity in this group, and the inverse of gN in the quotient group is $(gN)^{-1} = (g^{-1})N$.

Group table of the quotient group D_4 / R		
•	**R**	**U**
R	*R*	*U*
U	*U*	*R*

The elements of the quotient group D_4 / R are R itself, which represents the identity, and $U = f_vR$. The group operation on the quotient is shown at the right. For example, $U \cdot U = f_vR \cdot f_vR = (f_v \cdot f_v)R = R$. Both the subgroup $R = \{\text{id}, r_1, r_2, r_3\}$, as well as the corresponding quotient are abelian, whereas D_4 is not abelian. Building bigger groups by smaller ones, such as D_4 from its subgroup R and the quotient D_4 / R is abstracted by a notion called semidirect product.

Quotient groups and subgroups together form a way of describing every group by its *presentation*: any group is the quotient of the free group over the *generators* of the group, quotiented by the subgroup of *relations*. The dihedral group D_4, for example, can be generated by two elements r and f (for example, $r = r_1$, the right rotation and $f = f_v$ the vertical (or any other) reflection), which means that every symmetry of the square is a finite composition of these two symmetries or their inverses. Together with the relations

$$r^4 = f^2 = (r \bullet f)^2 = 1,$$

the group is completely described. A presentation of a group can also be used to construct the Cayley graph, a device used to graphically capture discrete groups.

Sub- and quotient groups are related in the following way: a subset H of G can be seen as an injective map $H \to G$, i.e., any element of the target has at most one element that maps to it. The counterpart to injective maps are surjective maps (every element of the target is mapped onto), such as the canonical map $G \to G / N$. Interpreting subgroup and quotients in light of these homomorphisms emphasizes the structural concept inherent to these definitions alluded to in the introduction. In general, homomorphisms are neither injective nor surjective. Kernel and image of group homomorphisms and the first isomorphism theorem address this phenomenon.

Examples and Applications

Examples and applications of groups abound. A starting point is the group Z of integers with addition as group operation, introduced above. If instead of addition multiplication is considered, one obtains multiplicative groups. These groups are predecessors of important constructions in abstract algebra.

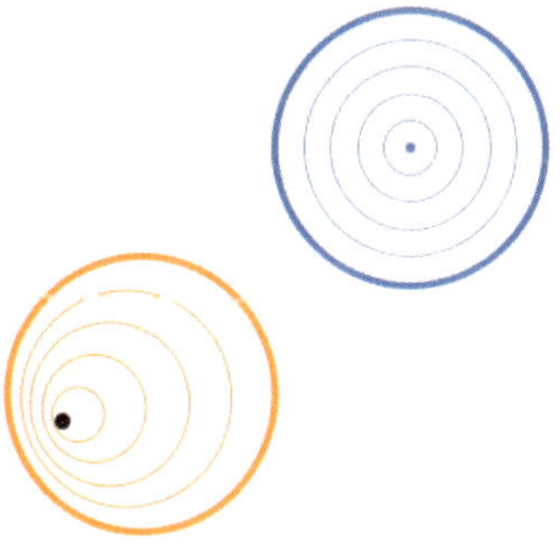

The fundamental group of a plane minus a point (bold) consists of loops around the missing point. This group is isomorphic to the integers.

Groups are also applied in many other mathematical areas. Mathematical objects are often examined by associating groups to them and studying the properties of the corresponding groups. For example, Henri Poincaré founded what is now called algebraic topology by introducing the fundamental group. By means of this connection, topological properties such as proximity and continuity translate into properties of groups. For example, elements of the fundamental group are represented by loops. The second image at the right shows some loops in a plane minus a point. The blue loop is considered null-homotopic (and thus irrelevant), because it can be continuously shrunk to a point. The presence of the hole prevents the orange loop from being shrunk to a point. The fundamental group of the plane with a point deleted turns out to be infinite cyclic, generated by the orange loop (or any other loop winding once around the hole). This way, the fundamental group detects the hole.

In more recent applications, the influence has also been reversed to motivate geometric constructions by a group-theoretical background. In a similar vein, geometric group theory employs geometric concepts, for example in the study of hyperbolic groups. Further branches crucially applying groups include algebraic geometry and number theory.

In addition to the above theoretical applications, many practical applications of groups exist. Cryptography relies on the combination of the abstract group theory approach together with algorithmical knowledge obtained in computational group theory, in particular when implemented for finite groups. Applications of group theory are not restricted to mathematics; sciences such as physics, chemistry and computer science benefit from the concept.

Numbers

Many number systems, such as the integers and the rationals enjoy a naturally given group structure. In some cases, such as with the rationals, both addition and multiplication operations give rise to group structures. Such number systems are predecessors to more general algebraic structures known as rings and fields. Further abstract algebraic concepts such as modules, vector spaces and algebras also form groups.

Integers

The group of integers Z under addition, denoted (Z, +), has been described above. The integers, with the operation of multiplication instead of addition, (Z, ·) do *not* form a group. The closure, associativity and identity axioms are satisfied, but inverses do not exist: for example, $a = 2$ is an integer, but the only solution to the equation $a \cdot b = 1$ in this case is $b = 1/2$, which is a rational number, but not an integer. Hence not every element of Z has a (multiplicative) inverse.

Rationals

The desire for the existence of multiplicative inverses suggests considering fractions

$$\frac{a}{b}.$$

Fractions of integers (with b nonzero) are known as rational numbers. The set of all such fractions is commonly denoted Q. There is still a minor obstacle for (Q, ·), the rationals with multiplication, being a group: because the rational number 0 does not have a multiplicative inverse (i.e., there is no x such that $x \cdot 0 = 1$), (Q, ·) is still not a group.

However, the set of all *nonzero* rational numbers $Q \setminus \{0\} = \{q \in Q \mid q \neq 0\}$ does form an abelian group under multiplication, denoted $(Q \setminus \{0\}, \cdot)$. Associativity and identity element axioms follow from the properties of integers. The closure requirement still holds true after removing zero, because the product of two nonzero rationals is never zero. Finally, the inverse of a/b is b/a, therefore the axiom of the inverse element is satisfied.

The rational numbers (including 0) also form a group under addition. Intertwining addition and multiplication operations yields more complicated structures called rings and—if division is possible, such as in Q—fields, which occupy a central position in abstract algebra. Group theoretic arguments therefore underlie parts of the theory of those entities.

Modular Arithmetic

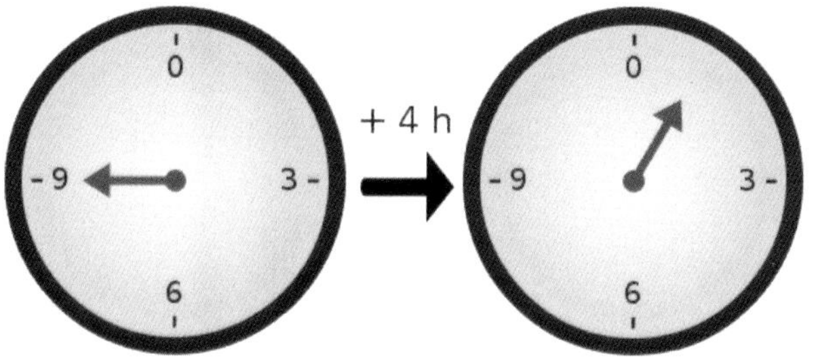

The hours on a clock form a group that uses addition modulo 12. Here 9 + 4 = 1.

In modular arithmetic, two integers are added and then the sum is divided by a positive integer called the *modulus*. The result of modular addition is the remainder of that division. For any modulus, n, the set of integers from 0 to $n - 1$ forms a group under modular addition: the inverse of any element a is $n - a$, and 0 is the identity element. This is familiar from the addition of hours on the face of a clock: if the hour hand is on 9 and is advanced 4 hours, it ends up on 1, as shown in the above. This is expressed by saying that 9 + 4 equals 1 "modulo 12" or, in symbols,

$$9 + 4 \equiv 1 \text{ modulo } 12.$$

The group of integers modulo n is written Z_n or Z/nZ.

For any prime number p, there is also the multiplicative group of integers modulo p. Its elements are the integers 1 to $p - 1$. The group operation is multiplication modulo p. That is, the usual product is divided by p and the remainder of this division is the result of modular multiplication. For example, if $p = 5$, there are four group elements 1, 2, 3, 4. In this group, $4 \cdot 4 = 1$, because the usual product 16 is equivalent to 1, which divided by 5 yields a remainder of 1. for 5 divides $16 - 1 = 15$, denoted

$$16 \equiv 1 \pmod 5.$$

The primality of p ensures that the product of two integers neither of which is divisible by p is not divisible by p either, hence the indicated set of classes is closed under multiplication. The identity element is 1, as usual for a multiplicative group, and the associativity follows from the corresponding property of integers. Finally, the inverse element axiom requires that given an integer a not divisible by p, there exists an integer b such that

$$a \cdot b \equiv 1 \pmod p\text{), i.e., } p \text{ divides the difference } a \cdot b - 1.$$

The inverse b can be found by using Bézout's identity and the fact that the greatest common divisor gcd(a, p) equals 1. In the case $p = 5$ above, the inverse of 4 is 4, and the inverse of 3 is 2, as $3 \cdot 2 = 6 \equiv 1 \pmod 5$. Hence all group axioms are fulfilled. Actually, this example is similar to $(Q \setminus \{0\}, \cdot)$ above: it consists of exactly those elements in Z/pZ that have a multiplicative inverse. These groups are denoted $F_p^\times$. They are crucial to public-key cryptography.

Cyclic Groups

A *cyclic group* is a group all of whose elements are powers of a particular element a. In multiplicative notation, the elements of the group are:

$$\ldots, a^{-3}, a^{-2}, a^{-1}, a^{0} = e, a, a^2, a^3, \ldots,$$

where a^2 means $a \bullet a$, and a^{-3} stands for $a^{-1} \bullet a^{-1} \bullet a^{-1} = (a \bullet a \bullet a)^{-1}$ etc. Such an element a is called a generator or a primitive element of the group. In additive notation, the requirement for an element to be primitive is that each element of the group can be written as

$$\ldots, -a-a, -a, 0, a, a+a, \ldots$$

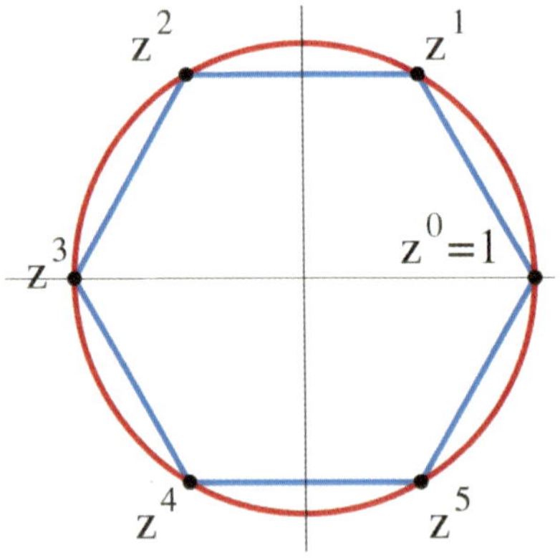

The 6th complex roots of unity form a cyclic group. z is a primitive element, but z^2 is not, because the odd powers of z are not a power of z^2.

In the groups Z/nZ introduced above, the element 1 is primitive, so these groups are cyclic. Indeed, each element is expressible as a sum all of whose terms are 1. Any cyclic group with n elements is isomorphic to this group. A second example for cyclic groups is the group of n-th complex roots of unity, given by complex numbers z satisfying $z^n = 1$. These numbers can be visualized as the vertices on a regular n-gon, as shown in blue for $n = 6$. The group operation is multiplication of complex numbers. In the picture, multiplying with z corresponds to a counter-clockwise rotation by 60°. Using some field theory, the group $F_p^{\times}$ can be shown to be cyclic: for example, if $p = 5$, 3 is a generator since $3^1 = 3$, $3^2 = 9 \equiv 4$, $3^3 \equiv 2$, and $3^4 \equiv 1$.

Some cyclic groups have an infinite number of elements. In these groups, for every non-zero element a, all the powers of a are distinct; despite the name "cyclic group", the powers of the elements do not cycle. An infinite cyclic group is isomorphic to (Z, +), the group of integers under addition introduced above. As these two prototypes are both abelian, so is any cyclic group.

The study of finitely generated abelian groups is quite mature, including the fundamental theorem of finitely generated abelian groups; and reflecting this state of affairs, many group-related notions, such as center and commutator, describe the extent to which a given group is not abelian.

Symmetry Groups

Symmetry groups are groups consisting of symmetries of given mathematical objects—be they of geometric nature, such as the introductory symmetry group of the square, or of algebraic nature, such as polynomial equations and their solutions. Conceptually, group theory can be thought of as the study of symmetry. Symmetries in mathematics greatly simplify the study of geometrical or analytical objects. A group is said to act on another mathematical object X if every group element performs some operation on X compatibly to the group law. In the rightmost example below, an element of order 7 of the (2,3,7) triangle group acts on the tiling by permuting the highlighted warped triangles (and the other ones, too). By a group action, the group pattern is connected to the structure of the object being acted on.

In chemical fields, such as crystallography, space groups and point groups describe molecular symmetries and crystal symmetries. These symmetries underlie the chemical and physical behavior of these systems, and group theory enables simplification of quantum mechanical analysis of these properties. For example, group theory is used to show that optical transitions between certain quantum levels cannot occur simply because of the symmetry of the states involved.

Not only are groups useful to assess the implications of symmetries in molecules, but surprisingly they also predict that molecules sometimes can change symmetry. The Jahn-Teller effect is a distortion of a molecule of high symmetry when it adopts a particular ground state of lower symmetry from a set of possible ground states that are related to each other by the symmetry operations of the molecule.

Likewise, group theory helps predict the changes in physical properties that occur when a material undergoes a phase transition, for example, from a cubic to a tetrahedral crystalline form. An example is ferroelectric materials, where the change from a paraelectric to a ferroelectric state occurs at the Curie temperature and is related to a change from the high-symmetry paraelectric state to the lower symmetry ferroelectric state, accompanied by a so-called soft phonon mode, a vibrational lattice mode that goes to zero frequency at the transition.

Such spontaneous symmetry breaking has found further application in elementary particle physics, where its occurrence is related to the appearance of Goldstone bosons.

Finite symmetry groups such as the Mathieu groups are used in coding theory, which is in turn applied in error correction of transmitted data, and in CD players. Another application is differential Galois theory, which characterizes functions having antiderivatives of a prescribed form, giving group-theoretic criteria for when solutions of certain differential equations are well-behaved. Geometric properties that remain stable under group actions are investigated in (geometric) invariant theory.

General Linear Group and Representation Theory

Matrix groups consist of matrices together with matrix multiplication. The *general linear group* GL(n, R) consists of all invertible n-by-n matrices with real entries. Its subgroups are referred to as *matrix groups* or *linear groups*. The dihedral group example mentioned above can be viewed as a (very small) matrix group. Another important matrix group is the special orthogonal group SO(n). It describes all possible rotations in n dimensions. Via Euler angles, rotation matrices are used in computer graphics.

Representation theory is both an application of the group concept and important for a deeper understanding of groups. It studies the group by its group actions on other spaces. A broad class of group representations are linear representations, i.e., the group is acting on a vector space, such as the three-dimensional Euclidean space R^3. A representation of G on an n-dimensional real vector space is simply a group homomorphism

$$\rho: G \to \mathrm{GL}(n, \mathrm{R})$$

from the group to the general linear group. This way, the group operation, which may be abstractly given, translates to the multiplication of matrices making it accessible to explicit computations.

Given a group action, this gives further means to study the object being acted on. On the other hand, it also yields information about the group. Group representations are an organizing principle in the theory of finite groups, Lie groups, algebraic groups and topological groups, especially (locally) compact groups.

Galois Groups

Galois groups were developed to help solve polynomial equations by capturing their symmetry features. For example, the solutions of the quadratic equation $ax^2 + bx + c = 0$ are given by

$$x = \frac{-b \pm \sqrt{b^2 - 4ac}}{2a}.$$

Exchanging "+" and "–" in the expression, i.e., permuting the two solutions of the equation can be viewed as a (very simple) group operation. Similar formulae are known for cubic and quartic equations, but do *not* exist in general for degree 5 and higher. Abstract properties of Galois groups associated with polynomials (in particular their solvability) give a criterion for polynomials that have all their solutions expressible by radicals, i.e., solutions expressible using solely addition, multiplication, and roots similar to the formula above.

The problem can be dealt with by shifting to field theory and considering the splitting field of a polynomial. Modern Galois theory generalizes the above type of Galois groups to field extensions and establishes—via the fundamental theorem of Galois theory—a precise relationship between fields and groups, underlining once again the ubiquity of groups in mathematics.

Finite Groups

A group is called *finite* if it has a finite number of elements. The number of elements is called the order of the group. An important class is the *symmetric groups* S_N, the groups of permutations of N letters. For example, the symmetric group on 3 letters S_3 is the group consisting of all possible orderings of the three letters *ABC*, i.e., contains the elements *ABC*, *ACB*, *BAC*, *BCA*, *CAB*, *CBA*, in total 6 (factorial of 3) elements. This class is fundamental insofar as any finite group can be expressed as a subgroup of a symmetric group S_N for a suitable integer N, according to Cayley's theorem. Parallel to the group of symmetries of the square above, S_3 can also be interpreted as the group of symmetries of an equilateral triangle.

The order of an element a in a group G is the least positive integer n such that $a^n = e$, where a^n represents

$$\underbrace{a \cdots a}_{n \text{ factors}},$$

i.e., application of the operation • to n copies of a. (If • represents multiplication, then a^n corresponds to the nth power of a.) In infinite groups, such an n may not exist, in which case the order of a is said to be infinity. The order of an element equals the order of the cyclic subgroup generated by this element.

More sophisticated counting techniques, for example counting cosets, yield more precise statements about finite groups: Lagrange's Theorem states that for a finite group G the order of any finite subgroup H divides the order of G. The Sylow theorems give a partial converse.

The dihedral group (discussed above) is a finite group of order 8. The order of r_1 is 4, as is the order of the subgroup R it generates. The order of the reflection elements f_v etc. is 2. Both orders divide 8, as predicted by Lagrange's theorem. The groups $F_p^\times$ above have order $p - 1$.

Classification of Finite Simple Groups

Mathematicians often strive for a complete classification (or list) of a mathematical notion. In the context of finite groups, this aim leads to difficult mathematics. According to Lagrange's theorem, finite groups of order p, a prime number, are necessarily cyclic (abelian) groups Z_p. Groups of order p^2 can also be shown to be abelian, a statement which does not generalize to order p^3, as the non-abelian group D_4 of order $8 = 2^3$ above shows. Computer algebra systems can be used to list small groups, but there is no classification of all finite groups. An intermediate step is the classification of finite simple groups. A nontrivial group is called *simple* if its only normal subgroups are the trivial group and the group itself. The Jordan–Hölder theorem exhibits finite simple groups as the building blocks for all finite groups. Listing all finite simple groups was a major achievement in contemporary group theory. 1998 Fields Medal winner Richard Borcherds succeeded in proving the monstrous moonshine conjectures, a surprising and deep relation between the largest finite simple sporadic group—the "monster group"—and certain modular functions, a piece of classical complex analysis, and string theory, a theory supposed to unify the description of many physical phenomena.

Groups with Additional Structure

Many groups are simultaneously groups and examples of other mathematical structures. In the language of category theory, they are group objects in a category, meaning that they are objects (that is, examples of another mathematical structure) which come with transformations (called morphisms) that mimic the group axioms. For example, every group (as defined above) is also a set, so a group is a group object in the category of sets.

Topological Groups

Some topological spaces may be endowed with a group law. In order for the group law and the topology to interweave well, the group operations must be continuous functions, that is, $g \cdot h$, and g^{-1} must not vary wildly if g and h vary only little. Such groups are called *topological groups,* and they are the group objects in the category of topological spaces. The most basic examples are the reals R under addition, (R \ {0}, ·), and similarly with any other topological field such as the complex numbers or p-adic numbers. All of these groups are locally compact, so they have Haar measures and can be studied via harmonic analysis. The former offer an abstract formalism of invariant integrals. Invariance means, in the case of real numbers for example:

$$\int_{\mathbb{R}} f(x)dx = \int_{\mathbb{R}} f(x+c)dx$$

for any constant c. Matrix groups over these fields fall under this regime, as do adele rings and adelic algebraic groups, which are basic to number theory. Galois groups of infinite field extensions such as the absolute Galois group can also be equipped with a topology, the so-called Krull topology, which in turn is central to generalize the above sketched connection of fields and groups to infinite field extensions. An advanced generalization of this idea, adapted to the needs of algebraic geometry, is the étale fundamental group.

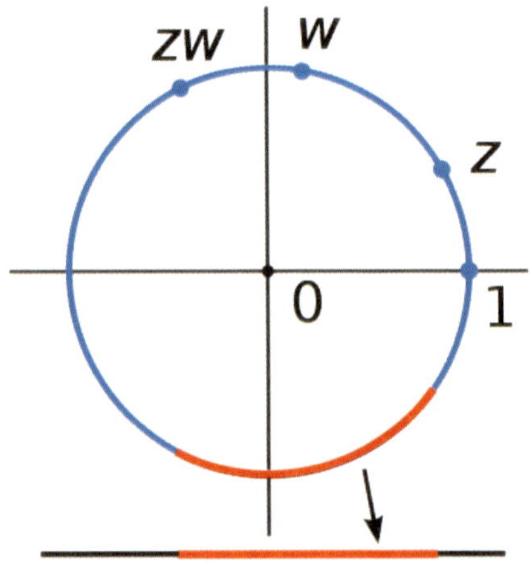

The unit circle in the complex plane under complex multiplication is a Lie group and, therefore, a topological group. It is topological since complex multiplication and division are continuous. It is a manifold and thus a Lie group, because every small piece, such as the red arc in the figure, looks like a part of the real line.

Lie Groups

Lie groups (in honor of Sophus Lie) are groups which also have a manifold structure, i.e., they are spaces looking locally like some Euclidean space of the appropriate dimension. Again, the additional structure, here the manifold structure, has to be compatible, i.e., the maps corresponding to multiplication and the inverse have to be smooth.

A standard example is the general linear group introduced above: it is an open subset of the space of all n-by-n matrices, because it is given by the inequality

$$\det(A) \neq 0,$$

where A denotes an n-by-n matrix.

Lie groups are of fundamental importance in modern physics: Noether's theorem links continuous symmetries to conserved quantities. Rotation, as well as translations in space and time are basic symmetries of the laws of mechanics. They can, for instance, be used to construct simple models—imposing, say, axial symmetry on a situation will typically lead to significant simplification in the equations one needs to solve to provide a physical description. Another example are the Lorentz transformations, which relate measurements of time and velocity of two observers in motion relative to each other. They can be deduced in a purely group-theoretical way, by expressing the transformations as a rotational symmetry of Minkowski space. The latter serves—in the absence of significant gravitation—as a model of space time in special relativity. The full symmetry group of Minkowski space, i.e., including translations, is known as the Poincaré group. By the above, it plays a pivotal role in special relativity and, by implication, for quantum field theories. Symmetries that vary with location are central to the modern description of physical interactions with the help of gauge theory.

Generalizations

Group-like structures					
	Totality[α]	Associativity	Identity	Invertibility	Commutativity
Semigroupoid	Unneeded	Required	Unneeded	Unneeded	Unneeded
Category	Unneeded	Required	Required	Unneeded	Unneeded
Groupoid	Unneeded	Required	Required	Required	Unneeded
Magma	Required	Unneeded	Unneeded	Unneeded	Unneeded
Quasigroup	Required	Unneeded	Unneeded	Required	Unneeded
Loop	Required	Unneeded	Required	Required	Unneeded
Semigroup	Required	Required	Unneeded	Unneeded	Unneeded
Monoid	Required	Required	Required	Unneeded	Unneeded
Group	Required	Required	Required	Required	Unneeded
Abelian group	Required	Required	Required	Required	Required

^**α** Closure, which is used in many sources, is an equivalent axiom to totality, though defined differently.

In abstract algebra, more general structures are defined by relaxing some of the axioms defining a group. For example, if the requirement that every element has an inverse is eliminated, the resulting algebraic structure is called a monoid. The natural numbers N (including 0) under addition form a monoid, as do the nonzero integers under multiplication (Z \ {0}, ·). There is a general method to formally add inverses to elements to any (abelian) monoid, much the same way as (Q \ {0}, ·) is derived from (Z \ {0}, ·), known as the Grothendieck group. Groupoids are similar to groups except that the composition $a \cdot b$ need not be defined for all a and b. They arise in the study of more complicated forms of symmetry, often in topological and analytical structures, such as the fundamental groupoid or stacks. Finally, it is possible to generalize any of these concepts by replacing the binary operation with an arbitrary n-ary one (i.e., an operation taking n arguments). With the proper generalization of the group axioms this gives rise to an n-ary group. The table gives a list of several structures generalizing groups.

Group Theory

The popular puzzle Rubik's cube invented by Ernő Rubik has been used as an illustration of permutation groups.

In mathematics and abstract algebra, group theory studies the algebraic structures known as groups. The concept of a group is central to abstract algebra: other well-known algebraic structures, such as rings, fields, and vector spaces, can all be seen as groups endowed with additional operations and axioms. Groups recur throughout mathematics, and the methods of group theory have influenced many parts of algebra. Linear algebraic groups and Lie groups are two branches of group theory that have experienced advances and have become subject areas in their own right.

Various physical systems, such as crystals and the hydrogen atom, may be modelled by symmetry groups. Thus group theory and the closely related representation theory have many important applications in physics, chemistry, and materials science. Group theory is also central to public key cryptography.

One of the most important mathematical achievements of the 20th century was the collaborative effort, taking up more than 10,000 journal pages and mostly published between 1960 and 1980, that culminated in a complete classification of finite simple groups.

Main Classes of Groups

The range of groups being considered has gradually expanded from finite permutation groups and special examples of matrix groups to abstract groups that may be specified through a presentation by generators and relations.

Permutation Groups

The first class of groups to undergo a systematic study was permutation groups. Given any set X and a collection G of bijections of X into itself (known as *permutations*) that is closed under compositions and inverses, G is a group acting on X. If X consists of n elements and G consists of *all* permutations, G is the symmetric group S_n; in general, any permutation group G is a subgroup of the symmetric group of X. An early construction due to Cayley exhibited any group as a permutation group, acting on itself ($X = G$) by means of the left regular representation.

In many cases, the structure of a permutation group can be studied using the properties of its action on the corresponding set. For example, in this way one proves that for $n \geq 5$, the alternating group A_n is simple, i.e. does not admit any proper normal subgroups. This fact plays a key role in the impossibility of solving a general algebraic equation of degree $n \geq 5$ in radicals.

Matrix Groups

The next important class of groups is given by *matrix groups*, or linear groups. Here G is a set consisting of invertible matrices of given order n over a field K that is closed under the products and inverses. Such a group acts on the n-dimensional vector space K^n by linear transformations. This action makes matrix groups conceptually similar to permutation groups, and the geometry of the action may be usefully exploited to establish properties of the group G.

Transformation Groups

Permutation groups and matrix groups are special cases of transformation groups: groups that act on a certain space X preserving its inherent structure. In the case of permutation groups, X is a set;

for matrix groups, X is a vector space. The concept of a transformation group is closely related with the concept of a symmetry group: transformation groups frequently consist of *all* transformations that preserve a certain structure.

The theory of transformation groups forms a bridge connecting group theory with differential geometry. A long line of research, originating with Lie and Klein, considers group actions on manifolds by homeomorphisms or diffeomorphisms. The groups themselves may be discrete or continuous.

Abstract Groups

Most groups considered in the first stage of the development of group theory were "concrete", having been realized through numbers, permutations, or matrices. It was not until the late nineteenth century that the idea of an abstract group as a set with operations satisfying a certain system of axioms began to take hold. A typical way of specifying an abstract group is through a presentation by *generators and relations*,

$$G = \langle S \mid R \rangle.$$

A significant source of abstract groups is given by the construction of a *factor group*, or quotient group, G/H, of a group G by a normal subgroup H. Class groups of algebraic number fields were among the earliest examples of factor groups, of much interest in number theory. If a group G is a permutation group on a set X, the factor group G/H is no longer acting on X; but the idea of an abstract group permits one not to worry about this discrepancy.

The change of perspective from concrete to abstract groups makes it natural to consider properties of groups that are independent of a particular realization, or in modern language, invariant under isomorphism, as well as the classes of group with a given such property: finite groups, periodic groups, simple groups, solvable groups, and so on. Rather than exploring properties of an individual group, one seeks to establish results that apply to a whole class of groups. The new paradigm was of paramount importance for the development of mathematics: it foreshadowed the creation of abstract algebra in the works of Hilbert, Emil Artin, Emmy Noether, and mathematicians of their school.

Topological and Algebraic Groups

An important elaboration of the concept of a group occurs if G is endowed with additional structure, notably, of a topological space, differentiable manifold, or algebraic variety. If the group operations m (multiplication) and i (inversion),

$$m: G \times G \to G, (g, h) \mapsto gh, \quad i: G \to G, g \mapsto g^{-1},$$

are compatible with this structure, i.e. are continuous, smooth or regular (in the sense of algebraic geometry) maps, then G becomes a topological group, a Lie group, or an algebraic group.

The presence of extra structure relates these types of groups with other mathematical disciplines and means that more tools are available in their study. Topological groups form a natural domain for abstract harmonic analysis, whereas Lie groups (frequently realized as transformation groups) are the mainstays of differential geometry and unitary representation theory. Certain classification questions that cannot be solved in general can be approached and resolved for special subclasses of groups. Thus, compact connected Lie groups have been completely classified. There is a fruitful relation between infinite abstract groups and topological groups: whenever a group Γ can be realized as a lattice in a topological group G, the geometry and analysis pertaining to G yield important results about Γ. A comparatively recent trend in the theory of finite groups exploits their connections with compact topological groups (profinite groups): for example, a single p-adic analytic group G has a family of quotients which are finite p-groups of various orders, and properties of G translate into the properties of its finite quotients.

Branches of Group Theory

Finite Group Theory

During the twentieth century, mathematicians investigated some aspects of the theory of finite groups in great depth, especially the local theory of finite groups and the theory of solvable and nilpotent groups. As a consequence, the complete classification of finite simple groups was achieved, meaning that all those simple groups from which all finite groups can be built are now known.

During the second half of the twentieth century, mathematicians such as Chevalley and Steinberg also increased our understanding of finite analogs of classical groups, and other related groups. One such family of groups is the family of general linear groups over finite fields. Finite groups often occur when considering symmetry of mathematical or physical objects, when those objects admit just a finite number of structure-preserving transformations. The theory of Lie groups, which may be viewed as dealing with "continuous symmetry", is strongly influenced by the associated Weyl groups. These are finite groups generated by reflections which act on a finite-dimensional Euclidean space. The properties of finite groups can thus play a role in subjects such as theoretical physics and chemistry.

Representation of Groups

Saying that a group G *acts* on a set X means that every element of G defines a bijective map on the set X in a way compatible with the group structure. When X has more structure, it is useful to restrict this notion further: a representation of G on a vector space V is a group homomorphism:

$$\rho : G \to \mathrm{GL}(V),$$

where GL(V) consists of the invertible linear transformations of V. In other words, to every group element g is assigned an automorphism $\rho(g)$ such that $\rho(g) \circ \rho(h) = \rho(gh)$ for any h in G.

This definition can be understood in two directions, both of which give rise to whole new domains of mathematics. On the one hand, it may yield new information about the group G: often, the group operation in G is abstractly given, but via ρ, it corresponds to the multiplication of matrices, which is very explicit. On the other hand, given a well-understood group acting on a complicated object, this simplifies the study of the object in question. For example, if G is finite, it is known that V above decomposes into irreducible parts. These parts in turn are much more easily manageable than the whole V (via Schur's lemma).

Given a group G, representation theory then asks what representations of G exist. There are several settings, and the employed methods and obtained results are rather different in every case: representation theory of finite groups and representations of Lie groups are two main subdomains of the theory. The totality of representations is governed by the group's characters. For example, Fourier polynomials can be interpreted as the characters of U(1), the group of complex numbers of absolute value 1, acting on the L^2-space of periodic functions.

Lie Theory

A Lie group is a group that is also a differentiable manifold, with the property that the group operations are compatible with the smooth structure. Lie groups are named after Sophus Lie, who laid the foundations of the theory of continuous transformation groups.

Lie groups represent the best-developed theory of continuous symmetry of mathematical objects and structures, which makes them indispensable tools for many parts of contemporary mathematics, as well as for modern theoretical physics. They provide a natural framework for analysing the continuous symmetries of differential equations (differential Galois theory), in much the same way as permutation groups are used in Galois theory for analysing the discrete symmetries of algebraic equations. An extension of Galois theory to the case of continuous symmetry groups was one of Lie's principal motivations.

Combinatorial and Geometric Group Theory

Groups can be described in different ways. Finite groups can be described by writing down the group table consisting of all possible multiplications $g \bullet h$. A more compact way of defining a group is by *generators and relations*, also called the *presentation* of a group. Given any set F of generators $\{g_i\}_{i \in I}$, the free group generated by F surjects onto the group G. The kernel of this map is called the subgroup of relations, generated by some subset D. The presentation is usually denoted by $\langle F \mid D \rangle$. For example, the group presentation $\langle a,b \mid aba^{-1}b^{-1} \rangle$ describes a group which is isomorphic to $\mathbb{Z} \times \mathbb{Z}$. A string consisting of generator symbols and their inverses is called a *word*.

Combinatorial group theory studies groups from the perspective of generators and relations. It is particularly useful where finiteness assumptions are satisfied, for example finitely generated groups, or finitely presented groups (i.e. in addition the relations are finite). The area makes use of the connection of graphs via their fundamental groups. For example, one can show that every subgroup of a free group is free.

There are several natural questions arising from giving a group by its presentation. The *word problem* asks whether two words are effectively the same group element. By relating the problem to Turing machines, one can show that there is in general no algorithm solving this task. Another, generally harder, algorithmically insoluble problem is the group isomorphism problem, which asks whether two groups given by different presentations are actually isomorphic. For example, the group with presentation $\langle x, y \mid xyxyx = e \rangle$, is isomorphic to the additive group Z of integers, although this may not be immediately apparent.

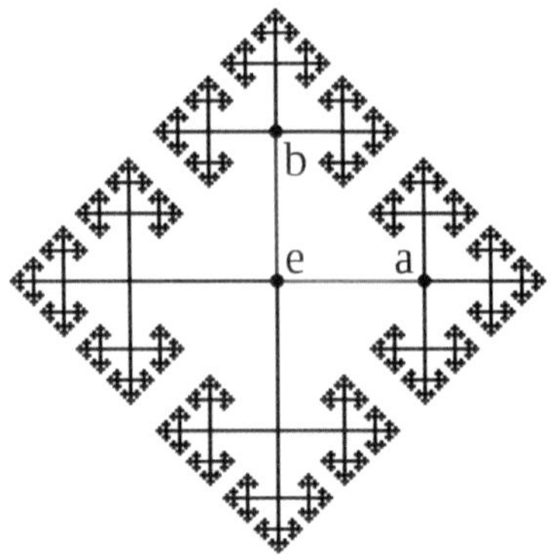

The Cayley graph of ⟨ x, y | ⟩, the free group of rank 2.

Geometric group theory attacks these problems from a geometric viewpoint, either by viewing groups as geometric objects, or by finding suitable geometric objects a group acts on. The first idea is made precise by means of the Cayley graph, whose vertices correspond to group elements and edges correspond to right multiplication in the group. Given two elements, one constructs the word metric given by the length of the minimal path between the elements. A theorem of Milnor and Svarc then says that given a group *G* acting in a reasonable manner on a metric space *X*, for example a compact manifold, then *G* is quasi-isometric (i.e. looks similar from a distance) to the space *X*.

Connection of Groups and Symmetry

Given a structured object *X* of any sort, a symmetry is a mapping of the object onto itself which preserves the structure. This occurs in many cases, for example

1. If *X* is a set with no additional structure, a symmetry is a bijective map from the set to itself, giving rise to permutation groups.
2. If the object *X* is a set of points in the plane with its metric structure or any other metric space, a symmetry is a bijection of the set to itself which preserves the distance between each pair of points (an isometry). The corresponding group is called isometry group of *X*.
3. If instead angles are preserved, one speaks of conformal maps. Conformal maps give rise to Kleinian groups, for example.
4. Symmetries are not restricted to geometrical objects, but include algebraic objects as well. For instance, the equation $x^2 - 3 = 0$ has the two solutions $\sqrt{3}$ and $-\sqrt{3}$. In this case, the group that exchanges the two roots is the Galois group belonging to the equation. Every polynomial equation in one variable has a Galois group, that is a certain permutation group on its roots.

The axioms of a group formalize the essential aspects of symmetry. Symmetries form a group: they are closed because if you take a symmetry of an object, and then apply another symmetry, the result will still be a symmetry. The identity keeping the object fixed is always a symmetry of an object. Existence of inverses is guaranteed by undoing the symmetry and the associativity comes from the fact that symmetries are functions on a space, and composition of functions are associative.

Frucht's theorem says that every group is the symmetry group of some graph. So every abstract group is actually the symmetries of some explicit object.

The saying of "preserving the structure" of an object can be made precise by working in a category. Maps preserving the structure are then the morphisms, and the symmetry group is the automorphism group of the object in question.

Applications of Group Theory

Applications of group theory abound. Almost all structures in abstract algebra are special cases of groups. Rings, for example, can be viewed as abelian groups (corresponding to addition) together with a second operation (corresponding to multiplication). Therefore, group theoretic arguments underlie large parts of the theory of those entities.

Galois Theory

Galois theory uses groups to describe the symmetries of the roots of a polynomial (or more precisely the automorphisms of the algebras generated by these roots). The fundamental theorem of Galois theory provides a link between algebraic field extensions and group theory. It gives an effective criterion for the solvability of polynomial equations in terms of the solvability of the corresponding Galois group. For example, S_5, the symmetric group in 5 elements, is not solvable which implies that the general quintic equation cannot be solved by radicals in the way equations of lower degree can. The theory, being one of the historical roots of group theory, is still fruitfully applied to yield new results in areas such as class field theory.

Algebraic Topology

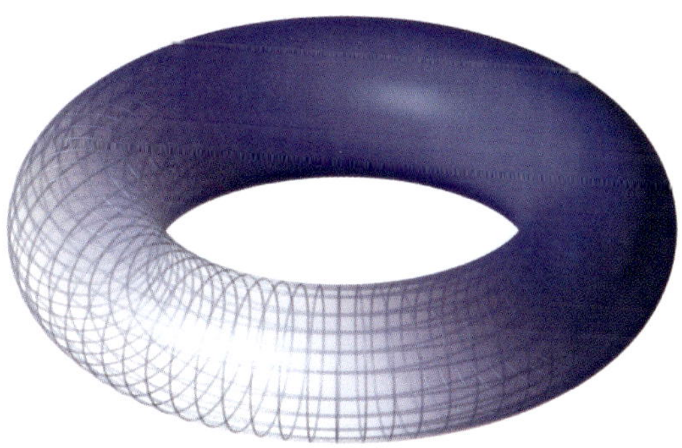

A torus. Its abelian group structure is induced from the map $C \to C/Z + \tau Z$, where τ is a parameter living in the upper half plane.

Algebraic topology is another domain which prominently associates groups to the objects the theory is interested in. There, groups are used to describe certain invariants of topological spaces. They are called "invariants" because they are defined in such a way that they do not change if the space is subjected to some deformation. For example, the fundamental group "counts" how many paths in the space are essentially different. The Poincaré conjecture, proved in 2002/2003 by Grigori Perelman, is a prominent application of this idea. The influence is not unidirectional, though. For example, algebraic topology makes use of Eilenberg–MacLane spaces which are spaces with prescribed homotopy groups. Similarly algebraic K-theory relies in a way on classifying spaces of groups. Finally, the name of the torsion subgroup of an infinite group shows the legacy of topology in group theory.

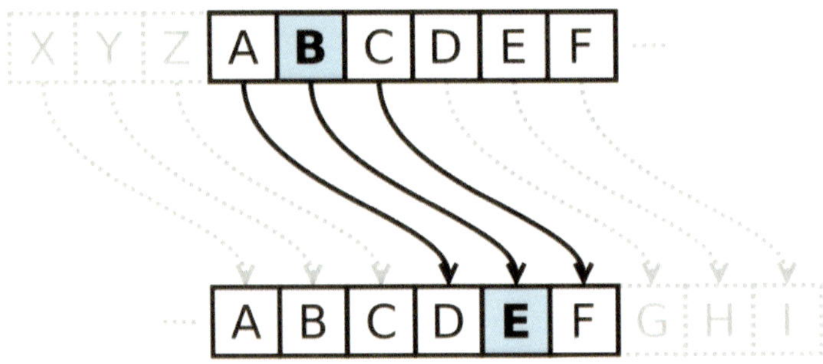

The cyclic group Z_{26} underlies Caesar's cipher.

Algebraic Geometry and Cryptography

Algebraic geometry and cryptography likewise uses group theory in many ways. Abelian varieties have been introduced above. The presence of the group operation yields additional information which makes these varieties particularly accessible. They also often serve as a test for new conjectures. The one-dimensional case, namely elliptic curves is studied in particular detail. They are both theoretically and practically intriguing. Very large groups of prime order constructed in elliptic curve cryptography serve for public-key cryptography. Cryptographical methods of this kind benefit from the flexibility of the geometric objects, hence their group structures, together with the complicated structure of these groups, which make the discrete logarithm very hard to calculate. One of the earliest encryption protocols, Caesar's cipher, may also be interpreted as a (very easy) group operation. In another direction, toric varieties are algebraic varieties acted on by a torus. Toroidal embeddings have recently led to advances in algebraic geometry, in particular resolution of singularities.

Algebraic Number Theory

Algebraic number theory is a special case of group theory, thereby following the rules of the latter. For example, Euler's product formula

$$\sum_{n\geq 1}\frac{1}{n^s}=\prod_{p\text{ prime}}\frac{1}{1-p^{-s}}$$

captures the fact that any integer decomposes in a unique way into primes. The failure of this statement for more general rings gives rise to class groups and regular primes, which feature in Kummer's treatment of Fermat's Last Theorem.

Harmonic Analysis

Analysis on Lie groups and certain other groups is called harmonic analysis. Haar measures, that is, integrals invariant under the translation in a Lie group, are used for pattern recognition and other image processing techniques.

Combinatorics

In combinatorics, the notion of permutation group and the concept of group action are often used to simplify the counting of a set of objects.

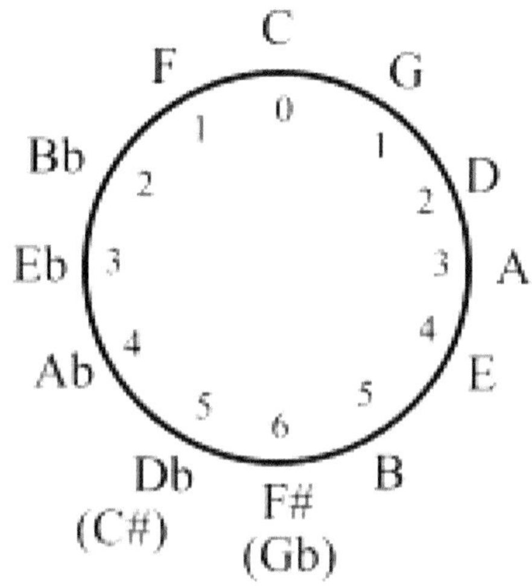

The circle of fifths may be endowed with a cyclic group structure

Music

The presence of the 12-periodicity in the circle of fifths yields applications of elementary group theory in musical set theory.

Physics

In physics, groups are important because they describe the symmetries which the laws of physics seem to obey. According to Noether's theorem, every continuous symmetry of a physical system corresponds to a conservation law of the system. Physicists are very interested in group representations, especially of Lie groups, since these representations often point the way to the "possible" physical theories. Examples of the use of groups in physics include the Standard Model, gauge theory, the Lorentz group, and the Poincaré group.

Chemistry and Materials Science

In chemistry and materials science, groups are used to classify crystal structures, regular polyhedra, and the symmetries of molecules. The assigned point groups can then be used to determine physical properties (such as chemical polarity and chirality), spectroscopic properties (particularly useful for Raman spectroscopy, infrared spectroscopy, circular dichroism spectroscopy, magnetic circular dichroism spectroscopy, UV/Vis spectroscopy, and fluorescence spectroscopy), and to construct molecular orbitals.

Molecular symmetry is responsible for many physical and spectroscopic properties of compounds and provides relevant information about how chemical reactions occur. In order to assign a point group for any given molecule, it is necessary to find the set of symmetry operations present on it. The symmetry operation is an action, such as a rotation around an axis or a reflection through a mirror plane. In other words, it is an operation that moves the molecule such that it is indistinguishable from the original configuration. In group theory, the rotation axes and mirror planes are called "symmetry elements". These elements can be a point, line or plane with respect to which the symmetry operation is carried out. The symmetry operations of a molecule determine the specific point group for this molecule.

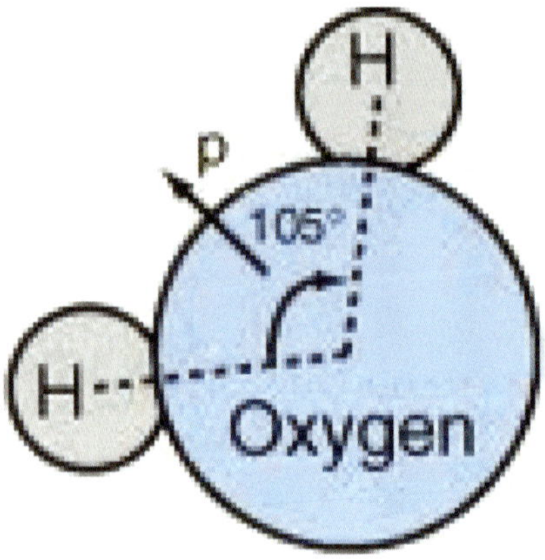

Water molecule with symmetry axis

In chemistry, there are five important symmetry operations. The identity operation (E) consists of leaving the molecule as it is. This is equivalent to any number of full rotations around any axis. This is a symmetry of all molecules, whereas the symmetry group of a chiral molecule consists of only the identity operation. Rotation around an axis (C_n) consists of rotating the molecule around a specific axis by a specific angle. For example, if a water molecule rotates 180° around the axis that passes through the oxygen atom and between the hydrogen atoms, it is in the same configuration as it started. In this case, $n = 2$, since applying it twice produces the identity operation. Other symmetry operations are: reflection, inversion and improper rotation (rotation followed by reflection).

Statistical Mechanics

Group theory can be used to resolve the incompleteness of the statistical interpretations of mechanics developed by Willard Gibbs, relating to the summing of an infinite number of probabilities to yield a meaningful solution.

Monoid

In abstract algebra, a branch of mathematics, a monoid is an algebraic structure with a single associative binary operation and an identity element.

Monoids are studied in semigroup theory, because they are semigroups with identity. Monoids occur in several branches of mathematics; for instance, they can be regarded as categories with a single object. Thus, they capture the idea of function composition within a set. In fact, all functions from a set into itself form naturally a monoid with respect to function composition. Monoids are also commonly used in computer science, both in its foundational aspects and in practical programming. The set of strings built from a given set of characters is a free monoid. The transition monoid and syntactic monoid are used in describing finite-state machines, whereas trace monoids and history monoids provide a foundation for process calculi and concurrent computing. Some of the more important results in the study of monoids are the Krohn–Rhodes theorem and the star height problem. The history of monoids, as well as a discussion of additional general properties, are found in the article on semigroups.

Definition

Suppose that S is a set and • is some binary operation $S \times S \rightarrow S$, then S with • is a monoid if it satisfies the following two axioms:

Associativity

For all a, b and c in S, the equation $(a \bullet b) \bullet c = a \bullet (b \bullet c)$ holds.

Identity Element

There exists an element e in S such that for every element a in S, the equations $e \bullet a = a \bullet e = a$ hold.

In other words, a monoid is a semigroup with an identity element. It can also be thought of as a magma with associativity and identity. The identity element of a monoid is unique. For this reason the identity is regarded as a constant, i. e. 0-ary (or nullary) operation. The monoid therefore is characterized by specification of the triple $(S, \bullet, e)$.

Depending on the context, the symbol for the binary operation may be omitted, so that the operation is denoted by juxtaposition; for example, the monoid axioms may be written $(ab)c = a(bc)$ and $ea = ae = a$. This notation does not imply that it is numbers being multiplied.

A monoid in which each element has an inverse is a group.

Monoid Structures

Submonoids

A submonoid of a monoid $(M, \bullet)$ is a subset N of M that is closed under the monoid operation and contains the identity element e of M. Symbolically, N is a submonoid of M if $N \subseteq M$, $x \bullet y \in N$ whenever $x, y \in N$, and $e \in N$. N is thus a monoid under the binary operation inherited from M.

Generators

A subset S of M is said to be a generator of M if M is the smallest set containing S that is closed under the monoid operation, or equivalently M is the result of applying the finitary closure operator to S. If there is a generator of M that has finite cardinality, then M is said to be finitely generated. Not every set S will generate a monoid, as the generated structure may lack an identity element.

Commutative Monoid

A monoid whose operation is commutative is called a commutative monoid (or, less commonly, an abelian monoid). Commutative monoids are often written additively. Any commutative monoid is endowed with its algebraic preordering $\leq$, defined by $x \leq y$ if there exists z such that $x + z = y$. An order-unit of a commutative monoid M is an element u of M such that for any element x of M, there exists a positive integer n such that $x \leq nu$. This is often used in case M is the positive cone of a partially ordered abelian group G, in which case we say that u is an order-unit of G.

Partially Commutative Monoid

A monoid for which the operation is commutative for some, but not all elements is a trace monoid; trace monoids commonly occur in the theory of concurrent computation.

Examples

- Out of the 16 possible binary Boolean operators, each of the four that has a two sided identity is also commutative and associative and thus makes the set {False, True} a commutative monoid. Under the standard definitions, AND and XNOR have the identity True while XOR and OR have the identity False. The monoids from AND and OR are also idempotent while those from XOR and XNOR are not.
- The natural numbers, N, form a commutative monoid under addition (identity element zero), or multiplication (identity element one). A submonoid of N under addition is called a numerical monoid.
- The positive integers, N \ {0}, form a commutative monoid under multiplication (identity element one).
- Given a set A, all subsets of A form a commutative monoid under intersection operation (identity element is A itself).
- Given a set A, all subsets of A form a commutative monoid under union operation (identity element is the empty set).
- Generalizing the previous example, every bounded semilattice is an idempotent commutative monoid.
    - In particular, any bounded lattice can be endowed with both a meet- and a join- monoid structure. The identity elements are the lattice's top and its bottom, respectively. Being lattices, Heyting algebras and Boolean algebras are endowed with these monoid structures.
- Every singleton set $\{x\}$ closed under a binary operation • forms the trivial (one-element) monoid, which is also the trivial group.
- Every group is a monoid and every abelian group a commutative monoid.
- Any semigroup S may be turned into a monoid simply by adjoining an element e not in S and defining $e \bullet s = s = s \bullet e$ for all $s \in S$. This conversion of any semigroup to the monoid is done by the free functor between the category of semigroups and the category of monoids.
    - Thus, an idempotent monoid (sometimes known as *find-first*) may be formed by adjoining an identity element e to the left zero semigroup over a set S. The opposite monoid (sometimes called *find-last*) is formed from the right zero semigroup over S.
        - Adjoin an identity e to the left-zero semigroup with two elements $\{lt; gt\}$. Then the resulting idempotent monoid $\{lt; e; gt\}$ models the lexicographical order of a sequence given the orders of its elements, with e representing equality.

- The elements of any unital ring, with addition or multiplication as the operation.
  - The integers, rational numbers, real numbers or complex numbers, with addition or multiplication as operation.
  - The set of all n by n matrices over a given ring, with matrix addition or matrix multiplication as the operation.
- The set of all finite strings over some fixed alphabet Σ forms a monoid with string concatenation as the operation. The empty string serves as the identity element. This monoid is denoted Σ^* and is called the free monoid over Σ.
- Given any monoid M, the *opposite monoid* M^{op} has the same carrier set and identity element as M, and its operation is defined by $x \bullet^{op} y = y \bullet x$. Any commutative monoid is the opposite monoid of itself.
- Given two sets M and N endowed with monoid structure (or, in general, any finite number of monoids, $M_1, \ldots, M_k$), their cartesian product $M \times N$ is also a monoid (respectively, $M_1 \times \ldots \times M_k$). The associative operation and the identity element are defined pairwise.
- Fix a monoid M. The set of all functions from a given set to M is also a monoid. The identity element is a constant function mapping any value to the identity of M; the associative operation is defined pointwise.
- Fix a monoid M with the operation • and identity element e, and consider its power set $P(M)$ consisting of all subsets of M. A binary operation for such subsets can be defined by $S \bullet T = \{ s \bullet t : s \in S, t \in T \}$. This turns $P(M)$ into a monoid with identity element $\{e\}$. In the same way the power set of a group G is a monoid under the product of group subsets.
- Let S be a set. The set of all functions $S \to S$ forms a monoid under function composition. The identity is just the identity function. It is also called the full transformation monoid of S. If S is finite with n elements, the monoid of functions on S is finite with n^n elements.
- Generalizing the previous example, let C be a category and X an object in C. The set of all endomorphisms of X, denoted $\mathrm{End}_C(X)$, forms a monoid under composition of morphisms. For more on the relationship between category theory and monoids see below.
- The set of homeomorphism classes of compact surfaces with the connected sum. Its unit element is the class of the ordinary 2-sphere. Furthermore, if a denotes the class of the torus, and b denotes the class of the projective plane, then every element c of the monoid has a unique expression the form $c = na + mb$ where n is the integer ≥ 0 and m = 0, 1, or 2. We have $3b = a + b$.
- Let $\langle f \rangle$ be a cyclic monoid of order n, that is, $\langle f \rangle = \{f^0, f^1, \ldots, f^{n-1}\}$. Then $f^n = f^k$ for some $0 \le k < n$. In fact, each such k gives a distinct monoid of order n, and every cyclic monoid is isomorphic to one of these.

Moreover, f can be considered as a function on the points $\{0,1,2,\ldots,n-1\}$ given by

$$\begin{bmatrix} 0 & 1 & 2 & \ldots & n-2 & n-1 \\ 1 & 2 & 3 & \ldots & n-1 & k \end{bmatrix}$$

or, equivalently

$$f(i) := \begin{cases} i+1, & \text{if } 0 \le i < n-1 \\ k, & \text{if } i = n-1. \end{cases}$$

Multiplication of elements in $\langle f \rangle$ is then given by function composition.

Note also that when $k = 0$ then the function f is a permutation of $\{0,1,2,\ldots,n-1\}$ and gives the unique cyclic group of order n.

Properties

In a monoid, one can define positive integer powers of an element x : $x^1 = x$, and $x^n = x \bullet \ldots \bullet x$ (n times) for $n > 1$. The rule of powers $x^{n+p} = x^n \bullet x^p$ is obvious.

From the definition of a monoid, one can show that the identity element e is unique. Then, for any x, one can set $x^0 = e$ and the rule of powers is still true with nonnegative exponents.

It is possible to define invertible elements: an element x is called invertible if there exists an element y such that $x \bullet y = e$ and $y \bullet x = e$. The element y is called the inverse of x. If y and z are inverses of x, then by associativity $y = (zx)y = z(xy) = z$. Thus inverses, if they exist, are unique.

If y is the inverse of x, one can define negative powers of x by setting $x^{-1} = y$ and $x^{-n} = y \bullet \ldots \bullet y$ (n times) for $n > 1$. And the rule of exponents is still verified for all integers n, p. This is why the inverse of x is usually written x^{-1}. The set of all invertible elements in a monoid M, together with the operation •, forms a group. In that sense, every monoid contains a group (possibly only the trivial group consisting of only the identity).

However, not every monoid sits inside a group. For instance, it is perfectly possible to have a monoid in which two elements a and b exist such that $a \bullet b = a$ holds even though b is not the identity element. Such a monoid cannot be embedded in a group, because in the group we could multiply both sides with the inverse of a and would get that $b = e$, which isn't true. A monoid $(M, \bullet)$ has the cancellation property (or is cancellative) if for all a, b and c in M, $a \bullet b = a \bullet c$ always implies $b = c$ and $b \bullet a = c \bullet a$ always implies $b = c$. A commutative monoid with the cancellation property can always be embedded in a group via the Grothendieck construction. That is how the additive group of the integers (a group with operation +) is constructed from the additive monoid of natural numbers (a commutative monoid with operation + and cancellation property). However, a non-commutative cancellative monoid need not be embeddable in a group.

If a monoid has the cancellation property and is *finite*, then it is in fact a group. Proof: Fix an element x in the monoid. Since the monoid is finite, $x^n = x^m$ for some $m > n > 0$. But then, by cancellation we have that $x^{m-n} = e$ where e is the identity. Therefore, $x \bullet x^{m-n-1} = e$, so x has an inverse.

The right- and left-cancellative elements of a monoid each in turn form a submonoid (i.e. obviously include the identity and not so obviously are closed under the operation). This means that the cancellative elements of any commutative monoid can be extended to a group.

It turns out that requiring the cancellative property in a monoid is not required to perform the Grothendieck construction – commutativity is sufficient. However, if the original monoid has an absorbing element then its Grothendieck group is the trivial group. Hence the homomorphism is, in general, not injective.

An inverse monoid is a monoid where for every a in M, there exists a unique a^{-1} in M such that $a = a \bullet a^{-1} \bullet a$ and $a^{-1} = a^{-1} \bullet a \bullet a^{-1}$. If an inverse monoid is cancellative, then it is a group.

In the opposite direction, a zerosumfree monoid is an additively written monoid in which $a + b = 0$ implies that $a = 0$ and $b = 0$: equivalently, that no element other than zero has an additive inverse.

Acts and Operator Monoids

Let M be a monoid, with the binary operation denoted by • and the identity element denoted by e. Then a (left) M-act (or left act over M) is a set X together with an operation $\cdot : M \times X \to X$ which is compatible with the monoid structure as follows:

- for all x in X: $e \cdot x = x$;
- for all a, b in M and x in X: $a \cdot (b \cdot x) = (a \bullet b) \cdot x$.

This is the analogue in monoid theory of a (left) group action. Right M-acts are defined in a similar way. A monoid with an act is also known as an operator monoid. Important examples include transition systems of semiautomata. A transformation semigroup can be made into an operator monoid by adjoining the identity transformation.

Monoid Homomorphisms

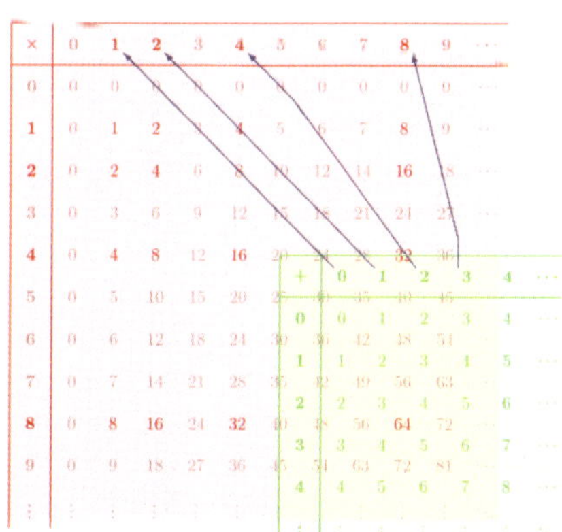

Example monoid homomorphism $x \mapsto 2^x$ from (N, +, 0) to (N, ×, 1). It is injective, but not surjective.

A homomorphism between two monoids $(M, *)$ and $(N, \bullet)$ is a function $f : M \to N$ such that

- $f(x * y) = f(x) \bullet f(y)$ for all x, y in M
- $f(e_M) = e_N$,

where e_M and e_N are the identities on M and N respectively. Monoid homomorphisms are sometimes simply called monoid morphisms.

Not every semigroup homomorphism is a monoid homomorphism, since it may not map the identity to the identity of the target monoid, even though the element it maps the identity to will be an identity of the image of the mapping. In contrast, a semigroup homomorphism between groups is always a group homomorphism, as it necessarily preserves the identity. Since for monoids this isn't always true, it is necessary to state this as a separate requirement.

A bijective monoid homomorphism is called a monoid isomorphism. Two monoids are said to be isomorphic if there is a monoid isomorphism between them.

Equational Presentation

Monoids may be given a presentation, much in the same way that groups can be specified by means of a group presentation. One does this by specifying a set of generators Σ, and a set of relations on the free monoid Σ^*. One does this by extending (finite) binary relations on Σ^* to monoid congruences, and then constructing the quotient monoid, as above.

Given a binary relation $R \subset \Sigma^* \times \Sigma^*$, one defines its symmetric closure as $R \cup R^{-1}$. This can be extended to a symmetric relation $E \subset \Sigma^* \times \Sigma^*$ by defining $x \sim_E y$ if and only if $x = sut$ and $y = svt$ for some strings $u, v, s, t \in \Sigma^*$ with $(u,v) \in R \cup R^{-1}$. Finally, one takes the reflexive and transitive closure of E, which is then a monoid congruence.

In the typical situation, the relation R is simply given as a set of equations, so that $R = \{u_1 = v_1, \cdots, u_n = v_n\}$. Thus, for example,

$$\langle p, q \mid pq = 1 \rangle$$

is the equational presentation for the bicyclic monoid, and

$$\langle a, b \mid aba = baa, bba = bab \rangle$$

is the plactic monoid of degree 2 (it has infinite order). Elements of this plactic monoid may be written as $a^i b^j (ba)^k$ for integers i, j, k, as the relations show that ba commutes with both a and b.

Relation to Category Theory

Group-like structures					
	Totality[α]	Associativity	Identity	Invertibility	Commutativity
Semigroupoid	Unneeded	Required	Unneeded	Unneeded	Unneeded
Category	Unneeded	Required	Required	Unneeded	Unneeded
Groupoid	Unneeded	Required	Required	Required	Unneeded
Magma	Required	Unneeded	Unneeded	Unneeded	Unneeded
Quasigroup	Required	Unneeded	Unneeded	Required	Unneeded
Loop	Required	Unneeded	Required	Required	Unneeded
Semigroup	Required	Required	Unneeded	Unneeded	Unneeded
Monoid	Required	Required	Required	Unneeded	Unneeded
Group	Required	Required	Required	Required	Unneeded
Abelian group	Required	Required	Required	Required	Required
^**α** Closure, which is used in many sources, is an equivalent axiom to totality, though defined differently.					

Monoids can be viewed as a special class of categories. Indeed, the axioms required of a monoid operation are exactly those required of morphism composition when restricted to the set of all morphisms whose source and target is a given object. That is,

A monoid is, essentially, the same thing as a category with a single object.

More precisely, given a monoid $(M, \bullet)$, one can construct a small category with only one object and whose morphisms are the elements of M. The composition of morphisms is given by the monoid operation $\bullet$.

Likewise, monoid homomorphisms are just functors between single object categories. So this construction gives an equivalence between the category of (small) monoids Mon and a full subcategory of the category of (small) categories Cat. Similarly, the category of groups is equivalent to another full subcategory of Cat.

In this sense, category theory can be thought of as an extension of the concept of a monoid. Many definitions and theorems about monoids can be generalised to small categories with more than one object. For example, a quotient of a category with one object is just a quotient monoid.

Monoids, just like other algebraic structures, also form their own category, Mon, whose objects are monoids and whose morphisms are monoid homomorphisms.

There is also a notion of monoid object which is an abstract definition of what is a monoid in a category. A monoid object in Set is just a monoid.

Monoids in Computer Science

In computer science, many abstract data types can be endowed with a monoid structure. In a common pattern, a sequence of elements of a monoid is "folded" or "accumulated" to produce a final value. For instance, many iterative algorithms need to update some kind of "running total" at each iteration; this pattern may be elegantly expressed by a monoid operation. Alternatively, the associativity of monoid operations ensures that the operation can be parallelized by employing a prefix sum or similar algorithm, in order to utilize multiple cores or processors efficiently.

Given a sequence of values of type M with identity element ε and associative operation $*$, the *fold* operation is defined as follows:

$$\text{fold} : M^* \to M = l \mapsto \begin{cases} \varepsilon & \text{if } l = \text{nil} \\ m * \text{fold} l' & \text{if } l = \text{cons } ml' \end{cases}$$

In addition, any data structure can be 'folded' in a similar way, given a serialization of its elements. For instance, the result of "folding" a binary tree might differ depending on pre-order vs. post-order tree traversal.

MapReduce

An application of monoids in computer science is so-called MapReduce programming model. MapReduce, in computing, consists of two or three operations. Given a dataset, "Map" consists of mapping arbitrary data to elements of a specific monoid. "Reduce" consists of folding those elements, so that in the end we produce just one element.

For example, if we have a multiset, in a program it is represented as a map from elements to their numbers. Elements are called keys in this case. The number of distinct keys may be too big, and in this case the multiset is being sharded. To finalize reduction properly, the "Shuffling" stage regroups the data among the nodes. If we do not need this step, the whole Map/Reduce consists of mapping and reducing; both operation are parallelizable, the former due to its element-wise nature, the latter due to associativity of the monoid.

Complete Monoids

A complete monoid is a commutative monoid equipped with an infinitary sum operation Σ_I for any index set I such that:

$$\sum_{i\in\varnothing} m_i = 0; \quad \sum_{i\in\{j\}} m_i = m_j; \quad \sum_{i\in\{j,k\}} m_i = m_j + m_k \quad \text{for } j \neq k$$

and

$$\sum_{j\in J}\sum_{i\in I_j} m_i = \sum_{i\in I}(m_i) \text{ if} \bigcup_{j\in J} I_j = I \text{ and } I_j \cap I_{j'} = \varnothing \text{ for } j \neq j'$$

A continuous monoid is an ordered commutative monoid in which every directed set has a least upper bound compatible with the monoid operation:

$$a + \sup S = \sup(a + S).$$

These two concepts are closely related: a continuous monoid is a complete monoid in which the infinitary sum may be defined as

$$\sum_I a_i = \sup \sum_E a_i$$

where the supremum on the right runs over all finite subsets E of I and each sum on the right is a finite sum in the monoid.

Ring

In mathematics, a ring is one of the fundamental algebraic structures used in abstract algebra. It consists of a set equipped with two binary operations that generalize the arithmetic operations of addition and multiplication. Through this generalization, theorems from arithmetic are extended to non-numerical objects such as polynomials, series, matrices and functions.

A ring is an abelian group with a second binary operation that is associative, is distributive over the abelian group operation, and has an identity element (this last property is not required by some authors). By extension from the integers, the abelian group operation is called *addition* and the second binary operation is called *multiplication*.

Whether a ring is commutative or not (i.e., whether the order in which two elements are multiplied changes the result or not) has profound implications on its behavior as an abstract object. As a result, commutative ring theory, commonly known as commutative algebra, is a key topic in ring theory. Its development has been greatly influenced by problems and ideas occurring naturally in algebraic number theory and algebraic geometry. Examples of commutative rings include the set of integers equipped with the addition and multiplication operations, the set of polynomials equipped with their addition and multiplication, the coordinate ring of an affine algebraic variety, and the ring of integers of a number field. Examples of noncommutative rings include the ring of $n \times n$ real square matrices with $n \geq 2$, group rings in representation theory, operator algebras in functional analysis, rings of differential operators in the theory of differential operators, and the cohomology ring of a topological space in topology.

The conceptualization of rings began in the 1870s and was completed in the 1920s. Key contributors include Dedekind, Hilbert, Fraenkel, and Noether. Rings were first formalized as a generalization of Dedekind domains that occur in number theory, and of polynomial rings and rings of invariants that occur in algebraic geometry and invariant theory. Afterward, they also proved to be useful in other branches of mathematics such as geometry and mathematical analysis.

Definition and Illustration

The most familiar example of a ring is the set of all integers, $\mathbb{Z}$, consisting of the numbers

$$\ldots, -5, -4, -3, -2, -1, 0, 1, 2, 3, 4, 5, \ldots$$

The familiar properties for addition and multiplication of integers serve as a model for the axioms for rings.

Definition

A ring is a set R equipped with two binary operations + and · satisfying the following three sets of axioms, called the ring axioms

1. R is an abelian group under addition, meaning that:

- $(a + b) + c = a + (b + c)$ for all a, b, c in R (that is, + is associative).
- $a + b = b + a$ for all a, b in R (that is, + is commutative).
- There is an element 0 in R such that $a + 0 = a$ for all a in R (that is, 0 is the additive identity).
- For each a in R there exists $-a$ in R such that $a + (-a) = 0$ (that is, $-a$ is the additive inverse of a).

2. R is a monoid under multiplication, meaning that:

- $(a \cdot b) \cdot c = a \cdot (b \cdot c)$ for all a, b, c in R (that is, · is associative).

- There is an element 1 in R such that $a \cdot 1 = a$ and $1 \cdot a = a$ for all a in R (that is, 1 is the multiplicative identity).

3. Multiplication is distributive with respect to addition:

   - $a \cdot (b + c) = (a \cdot b) + (a \cdot c)$ for all a, b, c in R (left distributivity).
   - $(b + c) \cdot a = (b \cdot a) + (c \cdot a)$ for all a, b, c in R (right distributivity).

Notes on the Definition

Many authors follow an alternative convention in which a ring is not defined to have a multiplicative identity. This article adopts the convention that, unless otherwise stated, a ring is assumed to have such an identity. A structure satisfying all the axioms *except* the requirement that there exists a multiplicative identity element is called a rng (commonly pronounced "rung", and sometimes called a pseudo-ring). For example, the set of even integers with the usual + and · is a rng, but not a ring.

The operations + and · are called *addition* and *multiplication*, respectively. The multiplication symbol · is often omitted, so the juxtaposition of ring elements is interpreted as multiplication. For example, xy means $x \cdot y$.

Although ring addition is commutative, ring multiplication is not required to be commutative: ab need not necessarily equal ba. Rings that also satisfy commutativity for multiplication (such as the ring of integers) are called commutative rings. Books on commutative algebra or algebraic geometry often adopt the convention that "ring" means "commutative ring", to simplify terminology.

In a ring, multiplication does not have to have an inverse. A commutative ring such that every nonzero element has a multiplicative inverse is called a field.

The additive group of a ring is the ring equipped just with the structure of addition. Although the definition assumes that the additive group is abelian, this can be inferred from the other ring axioms.

Basic Properties

Some basic properties of a ring follow immediately from the axioms:

- The additive identity, the additive inverse of each element, and the multiplicative identity are unique.
- For any element x in a ring R, one has $x0 = 0 = 0x$ (zero is an absorbing element with respect to multiplication) and $(-1)x = -x$.
- If $0 = 1$ in a ring R (or more generally, 0 is a unit element), then R has only one element, and is called the zero ring.
- The binomial formula holds for any commuting pair of elements (i.e., any x and y such that $xy = yx$).

Example: Integers Modulo 4

Equip the set $\mathbf{Z}_4 = \{\overline{0}, \overline{1}, \overline{2}, \overline{3}\}$ with the following operations:

- The sum $\overline{x} + \overline{y}$ in Z_4 is the remainder when the integer $x + y$ is divided by 4 (as $x + y$ is always smaller than 8, this remainder is either $x + y$ or $x + y - 4$). For example, $\overline{2} + \overline{3} = \overline{1}$ and $\overline{3} + \overline{3} = \overline{2}$.
- The product $\overline{x} \cdot \overline{y}$ in Z_4 is the remainder when the integer xy is divided by 4. For example, $\overline{2} \cdot \overline{3} = \overline{2}$ and $\overline{3} \cdot \overline{3} = \overline{1}$.

Then Z_4 is a ring: each axiom follows from the corresponding axiom for Z. If x is an integer, the remainder of x when divided by 4 may be considered as an element of Z_4, and this element is often denoted by "x mod 4" or $\overline{x}$, which is consistent with the notation for 0,1,2,3. The additive inverse of any $\overline{x}$ in Z_4 is $\overline{-x}$. For example, $-\overline{3} = \overline{-3} = \overline{1}$.

Example: 2-by-2 Matrices

The set of 2-by-2 matrices with real number entries is written

$$\mathcal{M}_2(\mathbb{R}) = \left\{ \begin{pmatrix} a & b \\ c & d \end{pmatrix} \middle| \, a, b, c, d \in \mathbb{R} \right\}.$$

With the operations of matrix addition and matrix multiplication, this set satisfies the above ring axioms. The element $\begin{pmatrix} 1 & 0 \\ 0 & 1 \end{pmatrix}$ is the multiplicative identity of the ring. If $A = \begin{pmatrix} 0 & 1 \\ 1 & 0 \end{pmatrix}$ and $B = \begin{pmatrix} 0 & 1 \\ 0 & 0 \end{pmatrix}$, then $AB = \begin{pmatrix} 0 & 0 \\ 0 & 1 \end{pmatrix}$ while $BA = \begin{pmatrix} 1 & 0 \\ 0 & 0 \end{pmatrix}$; this example shows that the ring is noncommutative.

More generally, for any ring R, commutative or not, and any nonnegative integer n, one may form the ring of n-by-n matrices with entries in R.

Basic Examples

Commutative rings:

- The prototype example is the ring of integers with the two operations of addition and multiplication.
- The rational, real and complex numbers are commutative rings of a type called fields.
- An algebra over a ring is itself a ring. These are also modules. Some examples:
  - Any algebra over a field.
  - The polynomial ring $R[X]$ of polynomials over a ring R is itself a ring. A free module over R of infinite dimension.

- $\mathbf{Z}[c]$, the integers with an irrational number c adjoined. A free module of infinite dimension if c is a transcendental number, a free module of finite dimension if c is an algebraic integer.
- $\mathbf{Z}[1/n]$, the set of fractions whose denominators are a power of n (including negative ones). A non-free module.
- $\mathbf{Z}[1/10]$, the set of decimal fractions.
- $\mathbf{Z}[(1+\sqrt{d})/2]$, where d is a square-free integer of the form $4n+1$. A free module of rank two. Cf. Quadratic integers.
- $\mathbf{Z}[i]$, the Gaussian integers.
- $\mathbf{Z}[(1+\sqrt{-3})/2]$, the Eisenstein integers. Also their generalization, a Kummer ring.

- The set of all algebraic integers forms a ring. This follows for example from the fact that it is the integral closure of the ring of rational integers in the field of complex numbers. The rings in the three previous examples are subrings of this ring.
- The set of formal power series $R[[X_1, ..., X_n]]$ over a commutative ring R is a ring.
- If S is a set, then the power set of S becomes a ring if we define addition to be the symmetric difference of sets and multiplication to be intersection. This corresponds to a ring of sets and is an example of a Boolean ring.
- The set of all continuous real-valued functions defined on the real line forms a commutative ring. The operations are pointwise addition and multiplication of functions.
- Let X be a set and R a ring. Then the set of all functions from X to R forms a ring, which is commutative if R is commutative. The ring of continuous functions in the previous example is a subring of this ring if X is the real line and R is the field of real numbers.

Noncommutative Rings:

- For any ring R and any natural number n, the set of all square n-by-n matrices with entries from R, forms a ring with matrix addition and matrix multiplication as operations. For $n = 1$, this matrix ring is isomorphic to R itself. For $n > 1$ (and R not the zero ring), this matrix ring is noncommutative.
- If G is an abelian group, then the endomorphisms of G form a ring, the endomorphism ring End(G) of G. The operations in this ring are addition and composition of endomorphisms. More generally, if V is a left module over a ring R, then the set of all R-linear maps forms a ring, also called the endomorphism ring and denoted by $\text{End}_R(V)$.
- If G is a group and R is a ring, the group ring of G over R is a free module over R having G as basis. Multiplication is defined by the rules that the elements of G commute with the elements of R and multiply together as they do in the group G.
- Many rings that appear in analysis are noncommutative. For example, most Banach algebras are noncommutative.

Non-rings:

- The set of natural numbers N with the usual operations is not a ring, since (N, +) is not even a group (the elements are not all invertible with respect to addition). For instance, there is no natural number which can be added to 3 to get 0 as a result. There is a natural way to make it a ring by adding negative numbers to the set, thus obtaining the ring of integers. The natural numbers (including 0) form an algebraic structure known as a semiring (which has all of the properties of a ring except the additive inverse property).
- Let R be the set of all continuous functions on the real line that vanish outside a bounded interval depending on the function, with addition as usual but with multiplication defined as convolution:

  $$(f*g)(x)=\int_{-\infty}^{\infty} f(y)g(x-y)dy.$$

  Then R is a rng, but not a ring: the Dirac delta function has the property of a multiplicative identity, but it is not a function and hence is not an element of R.

Basic Concepts

Elements in a Ring

A left zero divisor of a ring R is a nonzero element a in the ring such that there exists a nonzero element b of R such that $ab=0$. A right zero divisor is defined similarly.

A nilpotent element is an element a such that $a^n=0$ for some $n>0$. One example of a nilpotent element is a nilpotent matrix. A nilpotent element in a nonzero ring is necessarily a zero divisor.

An idempotent e is an element such that $e^2=e$. One example of an idempotent element is a projection in linear algebra.

A unit is an element a having a multiplicative inverse; in this case the inverse is unique, and is denoted by a^{-1}. The set of units of a ring is a group under ring multiplication; this group is denoted by $R^\times$ or R^* or $U(R)$. For example, if R is the ring of all square matrices of size n over a field, then $R^\times$ consists of the set of all invertible matrices of size n, and is called the general linear group.

Subring

A subset S of R is said to be a subring if it can be regarded as a ring with the addition and the multiplication restricted from R to S. Equivalently, S is a subring if it is not empty, and for any x, y in S, xy, $x+y$ and $-x$ are in S. If all rings have been assumed, by convention, to have a multiplicative identity, then to be a subring one would also require S to share the same identity element as R. So if all rings have been assumed to have a multiplicative identity, then a proper ideal is not a subring.

For example, the ring Z of integers is a subring of the field of real numbers and also a subring of the ring of polynomials Z[X] (in both cases, Z contains 1, which is the multiplicative identity of the larger rings). On the other hand, the subset of even integers 2Z does not contain the identity element 1 and thus does not qualify as a subring of Z.

An intersection of subrings is a subring. The smallest subring containing a given subset *E* of *R* is called a subring generated by *E*. Such a subring exists since it is the intersection of all subrings containing *E*.

For a ring *R*, the smallest subring containing 1 is called the *characteristic subring* of *R*. It can be obtained by adding copies of 1 and −1 together many times in any mixture. It is possible that $n \cdot 1 = 1 + 1 + \ldots + 1$ (*n* times) can be zero. If *n* is the smallest positive integer such that this occurs, then *n* is called the *characteristic* of *R*. In some rings, $n \cdot 1$ is never zero for any positive integer *n*, and those rings are said to have *characteristic zero.*

Given a ring *R*, let $Z(R)$ denote the set of all elements *x* in *R* such that *x* commutes with every element in *R*: $xy = yx$ for any *y* in *R*. Then $Z(R)$ is a subring of *R*; called the center of *R*. More generally, given a subset *X* of *R*, let *S* be the set of all elements in *R* that commute with every element in *X*. Then *S* is a subring of *R*, called the centralizer (or commutant) of *X*. The center is the centralizer of the entire ring *R*. Elements or subsets of the center are said to be *central* in *R*; they generate a subring of the center.

Ideal

The definition of an ideal in a ring is analogous to that of normal subgroup in a group. But, in actuality, it plays a role of an idealized generalization of an element in a ring; hence, the name "ideal". Like elements of rings, the study of ideals is central to structural understanding of a ring.

Let *R* be a ring. A nonempty subset *I* of *R* is then said to be a left ideal in *R* if, for any *x*, *y* in *I* and *r* in *R*, $x + y$ and rx are in *I*. If RI denotes the span of *I* over *R*; i.e., the set of finite sums

$$r_1x_1 + \cdots + r_nx_n, \quad r_i \in R, \quad x_i \in I,$$

then *I* is a left ideal if $RI \subseteq I$. Similarly, *I* is said to be right ideal if $IR \subseteq I$. A subset *I* is said to be a two-sided ideal or simply ideal if it is both a left ideal and right ideal. A one-sided or two-sided ideal is then an additive subgroup of *R*. If *E* is a subset of *R*, then RE is a left ideal, called the left ideal generated by *E*; it is the smallest left ideal containing *E*. Similarly, one can consider the right ideal or the two-sided ideal generated by a subset of *R*.

If *x* is in *R*, then Rx and xR are left ideals and right ideals, respectively; they are called the principal left ideals and right ideals generated by *x*. The principal ideal RxR is written as (x). For example, the set of all positive and negative multiples of 2 along with 0 form an ideal of the integers, and this ideal is generated by the integer 2. In fact, every ideal of the ring of integers is principal.

Like a group, a ring is said to be simple if it is nonzero and it has no proper nonzero two-sided ideals. A commutative simple ring is precisely a field.

Rings are often studied with special conditions set upon their ideals. For example, a ring in which there is no strictly increasing infinite chain of left ideals is called a left Noetherian ring. A ring in which there is no strictly decreasing infinite chain of left ideals is called a left Artinian ring. It is a somewhat surprising fact that a left Artinian ring is left Noetherian (the Hopkins–Levitzki theorem). The integers, however, form a Noetherian ring which is not Artinian.

For commutative rings, the ideals generalize the classical notion of divisibility and decomposition of an integer into prime numbers in algebra. A proper ideal P of R is called a prime ideal if for any elements $x, y \in R$ we have that $xy \in P$ implies either $x \in P$ or $y \in P$. Equivalently, P is prime if for any ideals I, J we have that $IJ \subseteq P$ implies either $I \subseteq P$ or $J \subseteq P$. This latter formulation illustrates the idea of ideals as generalizations of elements.

Homomorphism

A homomorphism from a ring $(R, +, \cdot)$ to a ring $(S, ‡, *)$ is a function f from R to S that preserves the ring operations; namely, such that, for all a, b in R the following identities hold:

- $f(a + b) = f(a) ‡ f(b)$
- $f(a \cdot b) = f(a) * f(b)$
- $f(1_R) = 1_S$

If one is working with not necessarily unital rings, then the third condition is dropped.

A ring homomorphism is said to be an isomorphism if there exists an inverse homomorphism to f (i.e., a ring homomorphism which is an inverse function). Any bijective ring homomorphism is a ring isomorphism. Two rings R, S are said to be isomorphic if there is an isomorphism between them and in that case one writes $R \simeq S$. A ring homomorphism between the same ring is called an endomorphism and an isomorphism between the same ring an automorphism.

Examples:

- The function that maps each integer x to its remainder modulo 4 (a number in {0, 1, 2, 3}) is a homomorphism from the ring Z to the quotient ring Z/4Z ("quotient ring" is defined below).
- If u is a unit element in a ring R, then $R \rightarrow R, x \mapsto uxu^{-1}$ is a ring homomorphism, called an inner automorphism of R.
- Let R be a commutative ring of prime characteristic p. Then $x \mapsto x^p$ is a ring endomorphism of R called the Frobenius homomorphism.
- The Galois group of a field extension L / K is the set of all automorphisms of L whose restrictions to K are the identity.
- For any ring R, there are a unique ring homomorphism Z →R and a unique ring homomorphism R →0.
- An epimorphism (i.e., right-cancelable morphism) of rings need not be surjective. For example, the unique map $\mathbb{Z} \rightarrow \mathbb{Q}$ is an epimorphism.
- An algebra homomorphism from a k-algebra to the endomorphism algebra of a vector space over k is called a representation of the algebra.

Given a ring homomorphism $f : R \rightarrow S$, the set of all elements mapped to 0 by f is called the kernel of f. The kernel is a two-sided ideal of R. The image of f, on the other hand, is not always an ideal, but it is always a subring of S.

To give a ring homomorphism from a commutative ring R to a ring A with image contained in the center of A is the same as to give a structure of an algebra over R to A (in particular gives a structure of A-module).

Quotient Ring

The quotient ring of a ring, is analogous to the notion of a quotient group of a group. More formally, given a ring $(R, +, \cdot)$ and a two-sided ideal I of $(R, +, \cdot)$, the quotient ring (or factor ring) R/I is the set of cosets of I (with respect to the additive group of $(R, +, \cdot)$; i.e. cosets with respect to $(R, +)$) together with the operations:

$$(a + I) + (b + I) = (a + b) + I \text{ and}$$

$$(a + I)(b + I) = (ab) + I.$$

for every a, b in R.

Like the case of a quotient group, there is a canonical map $p: R \to R/I$ given by $x \mapsto x + I$. It is surjective and satisfies the universal property: if $f: R \to S$ is a ring homomorphism such that $f(I) = 0$, then there is a unique $\overline{f}: R/I \to S$ such that $f = \overline{f} \circ p$. In particular, taking I to be the kernel, one sees that the quotient ring $R/\ker f$ is isomorphic to the image of f; the fact known as the first isomorphism theorem. The last fact implies that actually *any* surjective ring homomorphism satisfies the universal property since the image of such a map is a quotient ring.

Module

The concept of a *module over a ring* generalizes the concept of a vector space (over a field) by generalizing from multiplication of vectors with elements of a field (scalar multiplication) to multiplication with elements of a ring. More precisely, given a ring R with 1, an R-module M is an abelian group equipped with an operation $R \times M \to M$ (associating an element of M to every pair of an element of R and an element of M) that satisfies certain axioms. This operation is commonly denoted multiplicatively and called multiplication. The axioms of modules are the following: for all a, b in R and all x, y in M, we have:

- M is an abelian group under addition.
- $a(x + y) = ax + ay$
- $(a + b)x = ax + bx$
- $1x = x$
- $(ab)x = a(bx)$

When the ring is noncommutative these axioms define *left modules*; *right modules* are defined similarly by writing xa instead of ax. This is not only a change of notation, as the last axiom of right modules (that is $x(ab) = (xa)b$) becomes $(ab)x = b(ax)$, if left multiplication (by ring elements) is used for a right module.

Basic examples of modules are ideals, including the ring itself.

Although similarly defined, the theory of modules is much more complicated than that of vector space, mainly, because, unlike vector spaces, modules are not characterized (up to an isomorphism) by a single invariant (the dimension of a vector space). In particular, not all modules have a basis.

The axioms of modules imply that $(-1)x = -x$, where the first minus denotes the additive inverse in the ring and the second minus the additive inverse in the module. Using this and denoting repeated addition by a multiplication by a positive integer allows identifying abelian groups with modules over the ring of integers.

Any ring homomorphism induces a structure of a module: if $f : R \to S$ is a ring homomorphism, then S is a left module over R by the multiplication: $rs = f(r)s$. If R is commutative or if $f(R)$ is contained in the center of S, the ring S is called a R-algebra. In particular, every ring is an algebra over the integers.

Constructions

Direct Product

Let R and S be rings. Then the product $R \times S$ can be equipped with the following natural ring structure:

- $(r_1, s_1) + (r_2, s_2) = (r_1 + r_2, s_1 + s_2)$
- $(r_1, s_1) \cdot (r_2, s_2) = (r_1 \cdot r_2, s_1 \cdot s_2)$

for every r_1, r_2 in R and s_1, s_2 in S. The ring $R \times S$ with the above operations of addition and multiplication and the multiplicative identity $(1,1)$ is called the direct product of R with S. The same construction also works for an arbitrary family of rings: if R_i are rings indexed by a set I, then $\prod_{i \in I} R_i$ is a ring with componentwise addition and multiplication.

Let R be a commutative ring and $\mathfrak{a}_1, \cdots, \mathfrak{a}_n$ be ideals such that $\mathfrak{a}_i + \mathfrak{a}_j = (1)$ whenever $i \neq j$. Then the Chinese remainder theorem says there is a canonical ring isomorphism:

$$R / \bigcap_{i=1}^{n} \mathfrak{a}_i \simeq \prod_{i=1}^{n} R / \mathfrak{a}_i, \qquad x \bmod \bigcap_{i=1}^{n} \mathfrak{a}_i \mapsto (x \bmod \mathfrak{a}_1, \ldots, x \bmod \mathfrak{a}_n).$$

A "finite" direct product may also be viewed as a direct sum of ideals. Namely, let $R_i, 1 \le i \le n$ be rings, $R_i \to R = \prod R_i$ the inclusions with the images $\mathfrak{a}_i$ (in particular $\mathfrak{a}_i$ are rings though not subrings). Then $\mathfrak{a}_i$ are ideals of R and

$$R = \mathfrak{a}_1 \oplus \cdots \oplus \mathfrak{a}_n, \quad \mathfrak{a}_i \mathfrak{a}_j = 0, i \neq j, \quad \mathfrak{a}_i^2 \subseteq \mathfrak{a}_i$$

as a direct sum of abelian groups (because for abelian groups finite products are the same as direct sums). Clearly the direct sum of such ideals also defines a product of rings that is isomorphic to R. Equivalently, the above can be done through central idempotents. Assume R has the above decomposition. Then we can write

$$1 = e_1 + \cdots + e_n, \quad e_i \in \mathfrak{a}_i.$$

By the conditions on $\mathfrak{a}_i$, one has that e_i are central idempotents and $e_ie_j = 0, i \neq j$ (orthogonal). Again, one can reverse the construction. Namely, if one is given a partition of 1 in orthogonal central idempotents, then let $\mathfrak{a}_i = Re_i$ which are two-sided ideals. If each e_i is not a sum of orthogonal central idempotents, then their direct sum is isomorphic to *R*.

An important application of an infinite direct product is the construction of a projective limit of rings. Another application is a restricted product of a family of rings (cf. adele ring).

Polynomial Ring

Given a symbol *t* (called a variable) and a commutative ring *R*, the set of polynomials

$$R[t] = \{a_n t^n + a_{n-1} t^{n-1} + \ldots + a_1 t + a_0 \mid n \geq 0, a_j \in R\}$$

forms a commutative ring with the usual addition and multiplication, containing *R* as a subring. It is called the polynomial ring over *R*. More generally, the set $R[t_1, \ldots, t_n]$ of all polynomials in variables $t_1, \ldots, t_n$ forms a commutative ring, containing $R[t_i]$ as subrings.

If *R* is an integral domain, then $R[t]$ is also an integral domain; its field of fractions is the field of rational functions. If *R* is a noetherian ring, then $R[t]$ is a noetherian ring. If *R* is a unique factorization domain, then $R[t]$ is a unique factorization domain. Finally, *R* is a field if and only if $R[t]$ is a principal ideal domain.

Let $R \subseteq S$ be commutative rings. Given an element *x* of *S*, one can consider the ring homomorphism

$$R[t] \to S, \quad f \mapsto f(x)$$

(i.e., the substitution). If $S=R[t]$ and $x=t$, then $f(t)=f$. Because of this, the polynomial *f* is often also denoted by $f(t)$. The image of the map $f \mapsto f(x)$ is denoted by $R[x]$; it is the same thing as the subring of *S* generated by *R* and *x*.

Example: $k[t^2, t^3]$ denotes the image of the homomorphism

$$k[x, y] \to k[t], f \mapsto f(t^2, t^3).$$

In other words, it is the subalgebra of $k[t]$. generated by t^2 and t^3.

Example: let *f* be a polynomial in one variable; i.e., an element in a polynomial ring *R*. Then $f(x+h)$ is an element in $R[h]$ and $f(x+h) - f(x)$ is divisible by *h* in that ring. The result of substituting zero to *h* in $(f(x+h) - f(x))/h$ is $f'(x)$, the derivative of *f* at *x*.

The substitution is a special case of the universal property of a polynomial ring. The property states: given a ring homomorphism $\phi : R \to S$ and an element *x* in *S* there exists a unique ring homomorphism $\overline{\phi} : R[t] \to S$ such that $\overline{\phi}(t) = x$ and $\overline{\phi}$ restricts to ϕ . For example, choosing a basis, a symmetric algebra satisfies the universal property and so is a polynomial ring.

To give an example, let S be the ring of all functions from R to itself; the addition and the multiplication are those of functions. Let x be the identity function. Each r in R defines a constant function, giving rise to the homomorphism $R \to S$. The universal property says that this map extends uniquely to

$$R[t] \to S, \quad f \mapsto \overline{f}$$

(t maps to x) where $\overline{f}$ is the polynomial function defined by f. The resulting map is injective if and only if R is infinite.

Given a non-constant monic polynomial f in $R[t]$, there exists a ring S containing R such that f is a product of linear factors in $S[t]$.

Let k bc an algebraically closed field. The Hilbert's Nullstellensatz (theorem of zeros) states that there is a natural one-to-one correspondence between the set of all prime ideals in $k[t_1, \ldots, t_n]$ and the set of closed subvarieties of k^n. In particular, many local problems in algebraic geometry may be attacked through the study of the generators of an ideal in a polynomial ring. (cf. Gröbner basis.)

There are some other related constructions. A formal power series ring $R[[t]]$ consists of formal power series

$$\sum_0^\infty a_i t^i, \quad a_i \in R$$

together with multiplication and addition that mimic those for convergent series. It contains as a subring. Note a formal power series ring does not have the universal property of a polynomial ring; a series may not converge after a substitution. The important advantage of a formal power series ring over a polynomial ring is that it is local (in fact, complete).

Matrix Ring and Endomorphism Ring

Let R be a ring (not necessarily commutative). The set of all square matrices of size n with entries in R forms a ring with the entry-wise addition and the usual matrix multiplication. It is called the matrix ring and is denoted by $M_n(R)$. Given a right R-module U, the set of all R-linear maps from U to itself forms a ring with addition that is of function and multiplication that is of composition of functions; it is called the endomorphism ring of U and is denoted by $\mathrm{End}_R(U)$.

As in linear algebra, a matrix ring may be canonically interpreted as an endomorphism ring: $\mathrm{End}_R(R^n) \simeq M_n(R)$. This is a special case of the following fact: If $f : \oplus_1^n U \to \oplus_1^n U$ is an R-linear map, then f may be written as a matrix with entries f_{ij} in $S = \mathrm{End}_R(U)$, resulting in the ring isomorphism:

$$\mathrm{End}_R(\oplus_1^n U) \to M_n(S), \quad f \mapsto (f_{ij}).$$

Any ring homomorphism $R \to S$ induces $M_n(R) \to M_n(S)$; in fact, any ring homomorphism between matrix rings arises in this way.

Schur's lemma says that if U is a simple right R-module, then $\mathrm{End}_R(U)$ is a division ring. If $U = \bigoplus_{i=1}^{r} U_i^{\oplus m_i}$ is a direct sum of m_i-copies of simple R-modules U_i, then

$$\mathrm{End}_R(U) \simeq \bigoplus_{1}^{r} M_{m_i}(\mathrm{End}_R(U_i)).$$

The Artin–Wedderburn theorem states any semisimple ring (cf. below) is of this form.

A ring R and the matrix ring $M_n(R)$ over it are Morita equivalent: the category of right modules of R is equivalent to the category of right modules over $M_n(R)$. In particular, two-sided ideals in R correspond in one-to-one to two-sided ideals in $M_n(R)$.

Examples:

- The automorphisms of the projective line over a ring are given by homographies from the 2 x 2 matrix ring.

Limits and Colimits of Rings

Let R_i be a sequence of rings such that R_i is a subring of R_{i+1} for all i. Then the union (or filtered colimit) of R_i is the ring $\varinjlim R_i$ defined as follows: it is the disjoint union of all R_i's modulo the equivalence relation $x \sim y$ if and only if $x = y$ in R_i for sufficiently large i.

Examples of colimits:

- A polynomial ring in infinitely many variables: $R[t_1, t_2, \cdots] = \varinjlim R[t_1, t_2, \cdots, t_m]$.
- The algebraic closure of finite fields of the same characteristic $\overline{\mathbf{F}}_p = \varinjlim \mathbf{F}_{p^m}$.
- The field of formal Laurent series over a field k: $k((t)) = \varinjlim t^{-m} k[[t]]$ (it is the field of fractions of the formal power series ring $k[[t]]$.)
- The function field of an algebraic variety over a field k is $\varinjlim k[U]$ where the limit runs over all the coordinate rings $k[U]$ of nonempty open subsets U (more succinctly it is the stalk of the structure sheaf at the generic point.)

Any commutative ring is the colimit of finitely generated subrings.

A projective limit (or a filtered limit) of rings is defined as follows. Suppose we're given a family of rings R_i, i running over positive integers, say, and ring homomorphisms $R_j \to R_i, j \geq i$ such that $R_i \to R_i$ are all the identities and $R_k \to R_j \to R_i$ is $R_k \to R_i$ whenever $k \geq j \geq i$. Then $\varprojlim R_i$ is the subring of $\prod R_i$ consisting of (x_n) such that x_j maps to x_i under $R_j \to R_i, j \geq i$.

For an example of a projective limit.

Localization

The localization generalizes the construction of the field of fractions of an integral domain to an arbitrary ring and modules. Given a (not necessarily commutative) ring R and a subset S of R, there exists a ring $R[S^{-1}]$ together with the ring homomorphism $R \to R[S^{-1}]$ that "inverts" S; that is, the homomorphism maps elements in S to unit elements in $R[S^{-1}]$, and, moreover, any ring homomorphism from R that "inverts" S uniquely factors through $R[S^{-1}]$. The ring $R[S^{-1}]$ is called the localization of R with respect to S. For example, if R is a commutative ring and f an element in R, then the localization $R[f^{-1}]$ consists of elements of the form $r / f^n, r \in R, n \geq 0$ (to be precise, $R[f^{-1}] = R[t]/(tf-1)$.)

The localization is frequently applied to a commutative ring R with respect to the complement of a prime ideal (or a union of prime ideals) in R. In that case $S = R - \mathfrak{p}$, one often writes $R_\mathfrak{p}$ for $R[S^{-1}]$. $R_\mathfrak{p}$ is then a local ring with the maximal ideal $\mathfrak{p}R_\mathfrak{p}$. This is the reason for the terminology "localization". The field of fractions of an integral domain R is the localization of R at the prime ideal zero. If $\mathfrak{p}$ is a prime ideal of a commutative ring R, then the field of fractions of $R/\mathfrak{p}$ is the same as the residue field of the local ring $R_\mathfrak{p}$ and is denoted by $k(\mathfrak{p})$.

If M is a left R-module, then the localization of M with respect to S is given by a change of rings $M[S^{-1}] = R[S^{-1}] \otimes_R M$.

The most important properties of localization are the following: when R is a commutative ring and S a multiplicatively closed subset

- $\mathfrak{p} \mapsto \mathfrak{p}[S^{-1}]$ is a bijection between the set of all prime ideals in R disjoint from S and the set of all prime ideals in $R[S^{-1}]$.
- $R[S^{-1}] = \varinjlim R[f^{-1}]$, f running over elements in S with partial ordering given by divisibility.
- The localization is exact:

  $0 \to M'[S^{-1}] \to M[S^{-1}] \to M''[S^{-1}] \to 0$ is exact over $R[S^{-1}]$ whenever

  $0 \to M' \to M \to M'' \to 0$ is exact over R.
- Conversely, if $0 \to M'_\mathfrak{m} \to M_\mathfrak{m} \to M''_\mathfrak{m} \to 0$ is exact for any maximal ideal $\mathfrak{m}$, then $0 \to M' \to M \to M'' \to 0$ is exact.
- A remark: localization is no help in proving a global existence. One instance of this is that if two modules are isomorphic at all prime ideals, it does not follow that they are isomorphic. (One way to explain this is that the localization allows one to view a module as a sheaf over prime ideals and a sheaf is inherently a local notion.)

In category theory, a localization of a category amounts to making some morphisms isomorphisms. An element in a commutative ring R may be thought of as an endomorphism of any R-module. Thus, categorically, a localization of R with respect to a subset S of R is a functor from the category of R-modules to itself that sends elements of S viewed as endomorphisms to automorphisms and is universal with respect to this property. (Of course, R then maps to $R[S^{-1}]$ and R-modules map to $R[S^{-1}]-$ modules.)

Completion

Let R be a commutative ring, and let I be an ideal of R. The completion of R at I is the projective limit $\hat{R} = \varprojlim R / I^n$; it is a commutative ring. The canonical homomorphisms from R to the quotients R / I^n induce a homomorphism $R \to \hat{R}$. The latter homomorphism is injective if R is a noetherian integral domain and I is a proper ideal, or if R is a noetherian local ring with maximal ideal I, by Krull's intersection theorem. The construction is especially useful when I is a maximal ideal.

The basic example is the completion Z_p of Z at the principal ideal (p) generated by a prime number p; it is called the ring of p-adic integers. The completion can in this case be constructed also from the p-adic absolute value on Q. The p-adic absolute value on Q is a map $x \mapsto |x|$ from Q to R given by $|n|_p = p^{-v_p(n)}$ where $v_p(n)$ denotes the exponent of p in the prime factorization of a nonzero integer n into prime numbers (we also put $|0|_p = 0$ and $|m/n|_p = |m|_p / |n|_p$). It defines a distance function on Q and the completion of Q as a metric space is denoted by Q_p. It is again a field since the field operations extend to the completion. The subring of Q_p consisting of elements x with $|x|_p \leq 1$ is isomorphic to Z_p.

Similarly, the formal power series ring $R[[t]]$ is the completion of $R[t]$ at (t)

A complete ring has much simpler structure than a commutative ring. This owns to the Cohen structure theorem, which says, roughly, that a complete local ring tends to look like a formal power series ring or a quotient of it. On the other hand, the interaction between the integral closure and completion has been among the most important aspects that distinguish modern commutative ring theory from the classical one developed by the likes of Noether. Pathological examples found by Nagata led to the reexamination of the roles of Noetherian rings and motivated, among other things, the definition of excellent ring.

Rings with Generators and Relations

The most general way to construct a ring is by specifying generators and relations. Let F be a free ring (i.e., free algebra over the integers) with the set X of symbols; i.e., F consists of polynomials with integral coefficients in noncommuting variables that are elements of X. A free ring satisfies the universal property: any function from the set X to a ring R factors through F so that $F \to R$ is the unique ring homomorphism. Just as in the group case, every ring can be represented as a quotient of a free ring.

Now, we can impose relations among symbols in X by taking a quotient. Explicitly, if E is a subset of F, then the quotient ring of F by the ideal generated by E is called the ring with generators X and relations E. If we used a ring, say, A as a base ring instead of Z, then the resulting ring will be over A. For example, if $E = \{xy - yx \mid x, y \in X\}$, then the resulting ring will be the usual polynomial ring with coefficients in A in variables that are elements of X (It is also the same thing as the symmetric algebra over A with symbols X.)

In the category-theoretic terms, the formation $S \mapsto$ the free ring generated by the set S is the left adjoint functor of the forgetful functor from the category of rings to Set (and it is often called the free ring functor.)

Let A, B be algebras over a commutative ring R. Then the tensor product of R-modules $A \otimes_R B$ is a R-module. We can turn it to a ring by extending linearly $(x \otimes u)(y \otimes v) = xy \otimes uv$.

Special Kinds of Rings

Domains

A nonzero ring with no nonzero zero-divisors is called a domain. A commutative domain is called an integral domain. The most important integral domains are principal ideals domains, PID for short, and fields. A principal ideal domain is an integral domain in which every ideal is principal. An important class of integral domains that contain a PID is a unique factorization domain (UFD), an integral domain in which every nonunit element is a product of prime elements (an element is prime if it generates a prime ideal.) The fundamental question in algebraic number theory is on the extent to which the ring of (generalized) integers in a number field, where an "ideal" admits prime factorization, fails to be a PID.

Among theorems concerning a PID, the most important one is the structure theorem for finitely generated modules over a principal ideal domain. The theorem may be illustrated by the following application to linear algebra. Let V be a finite-dimensional vector space over a field k and $f: V \to V$ a linear map with minimal polynomial q. Then, since $k[t]$ is a unique factorization domain, q factors into powers of distinct irreducible polynomials (i.e., prime elements):

$$q = p_1^{e_1} \ldots p_s^{e_s}.$$

Letting $t \cdot v = f(v)$, we make V a $k[t]$-module. The structure theorem then says V is a direct sum of cyclic modules, each of which is isomorphic to the module of the form $k[t]/(p_i^{k_j})$. Now, if $p_i(t) = t - \lambda_i$ then such a cyclic module (for p_i) has a basis in which the restriction of f is represented by a Jordan matrix. Thus, if, say, k is algebraically closed, then all p_i 's are of the form $t - \lambda_i$ and the above decomposition corresponds to the Jordan canonical form of f.

In algebraic geometry, UFDs arise because of smoothness. More precisely, a point in a variety (over a perfect field) is smooth if the local ring at the point is a regular local ring. A regular local ring is a UFD.

The following is a chain of class inclusions that describes the relationship between rings, domains and fields:

> Commutative rings ⊃ integral domains ⊃ integrally closed domains ⊃unique factorization domains ⊃ principal ideal domains ⊃ Euclidean domains ⊃ fields

Division Ring

A division ring is a ring such that every non-zero element is a unit. A commutative division ring is a field. A prominent example of a division ring that is not a field is the ring of quaternions. Any centralizer in a division ring is also a division ring. In particular, the center of a division ring is a field. It turned out that every *finite* domain (in particular finite division ring) is a field; in particular commutative (the Wedderburn's little theorem).

Every module over a division ring is a free module (has a basis); consequently, much of linear algebra can be carried out over a division ring instead of a field.

The study of conjugacy classes figures prominently in the classical theory of division rings. Cartan famously asked the following question: given a division ring D and a proper sub-division-ring S that is not contained in the center, does each inner automorphism of D restrict to an automorphism of S? The answer is negative: this is the Cartan–Brauer–Hua theorem.

A cyclic algebra, introduced by L. E. Dickson, is a generalization of a quaternion algebra.

Semisimple Rings

A ring is called a semisimple ring if it is semisimple as a left module (or right module) over itself; i.e., a direct sum of simple modules. A ring is called a semiprimitive ring if its Jacobson radical is zero. (The Jacobson radical is the intersection of all maximal left ideals.) A ring is semisimple if and only if it is artinian and is semiprimitive.

An algebra over a field k is artinian if and only if it has finite dimension. Thus, a semisimple algebra over a field is necessarily finite-dimensional, while a simple algebra may have infinite dimension; e.g., the ring of differential operators.

Any module over a semisimple ring is semisimple. (Proof: any free module over a semisimple ring is clearly semisimple and any module is a quotient of a free module.)

Examples of semisimple rings:

- A matrix ring over a division ring is semisimple (actually simple).
- The group ring $k[G]$ of a finite group G over a field k is semisimple if the characteristic of k does not divide the order of G. (Maschke's theorem)
- The Weyl algebra (over a field) is a simple ring; it is not semisimple since it has infinite dimension and thus not artinian.
- Clifford algebras are semisimple.

Semisimplicity is closely related to separability. An algebra A over a field k is said to be separable if the base extension $A \otimes_k F$ is semisimple for any field extension F / k. If A happens to be a field, then this is equivalent to the usual definition in field theory (cf. separable extension.)

Central Simple Algebra and Brauer Group

For a field k, a k-algebra is central if its center is k and is simple if it is a simple ring. Since the center of a simple k-algebra is a field, any simple k-algebra is a central simple algebra over its center. In this section, a central simple algebra is assumed to have finite dimension. Also, we mostly fix the base field; thus, an algebra refers to a k-algebra. The matrix ring of size n over a ring R will be denoted by R_n.

The Skolem–Noether theorem states any automorphism of a central simple algebra is inner.

Two central simple algebras A and B are said to be *similar* if there are integers n and m such that $A \otimes_k k_n \approx B \otimes_k k_m$. Since $k_n \otimes_k k_m \simeq k_{nm}$, the similarity is an equivalence relation. The similarity classes $[A]$ with the multiplication $[A][B]=[A\otimes_k B]$ form an abelian group called the Brauer group of *k* and is denoted by $\mathrm{Br}(k)$. By the Artin–Wedderburn theorem, a central simple algebra is the matrix ring of a division ring; thus, each similarity class is represented by a unique division ring.

For example, $\mathrm{Br}(k)$ is trivial if *k* is a finite field or an algebraically closed field (more generally quasi-algebraically closed field; cf. Tsen's theorem). $\mathrm{Br}(\mathbb{R})$ has order 2 (a special case of the theorem of Frobenius). Finally, if *k* is a nonarchimedean local field (e.g., $\mathbb{Q}_p$), then $\mathrm{Br}(k)=\mathbb{Q}/\mathbb{Z}$ through the invariant map.

Now, if F is a field extension of k, then the base extension $-\otimes_k F$ induces $\mathrm{Br}(k)\to\mathrm{Br}(F)$. Its kernel is denoted by $\mathrm{Br}(F/k)$. It consists of $[A]$ such that $A\otimes_k F$ is a matrix ring over *F* (i.e., *A* is split by *F*.) If the extension is finite and Galois, then $\mathrm{Br}(F/k)$ is canonically isomorphic to $H^2(\mathrm{Gal}(F/k),k^*)$.

Azumaya algebras generalize the notion of central simple algebras to a commutative local ring.

Valuation Ring

If *K* is a field, a valuation v is a group homomorphism from the multiplicative group K^* to a totally ordered abelian group *G* such that, for any *f*, *g* in *K* with $f + g$ nonzero, $v(f + g) \geq \min\{v(f), v(g)\}$. The valuation ring of v is the subring of *K* consisting of zero and all nonzero *f* such that $v(f) \geq 0$.

Examples:

- The field of formal Laurent series $k((t))$ over a field *k* comes with the valuation v such that $v(f)$ is the least degree of a nonzero term in *f*; the valuation ring of v is the formal power series ring $k[[t]]$.
- More generally, given a field *k* and a totally ordered abelian group *G*, let $k((G))$ be the set of all functions from *G* to *k* whose supports (the sets of points at which the functions are nonzero) are well ordered. It is a field with the multiplication given by convolution:

$$(f*g)(t)=\sum_{s\in G} f(s)g(t-s).$$

  It also comes with the valuation v such that $v(f)$ is the least element in the support of *f*. The subring consisting of elements with finite support is called the group ring of *G* (which makes sense even if *G* is not commutative). If *G* is the ring of integers, then we recover the previous example (by identifying *f* with the series whose *n*-th coefficient is $f(n)$.)

Rings with Extra Structure

A ring may be viewed as an abelian group (by using the addition operation), with extra structure: namely, ring multiplication. In the same way, there are other mathematical objects which may be considered as rings with extra structure. For example:

- An associative algebra is a ring that is also a vector space over a field *K* such that the scalar multiplication distributes over the ring multiplication. For instance, the set of *n*-by-*n* matrices over the real field R has dimension n^2 as a real vector space.

- A ring R is a topological ring if its set of elements R is given a topology which makes the addition map ($+: R \times R \to R$) and the multiplication map ($\cdot: R \times R \to R$) to be both continuous as maps between topological spaces (where $X \times X$ inherits the product topology or any other product in the category). For example, n-by-n matrices over the real numbers could be given either the Euclidean topology, or the Zariski topology, and in either case one would obtain a topological ring.
- A λ-ring is a commutative ring R together with operations $\lambda^n: R \to R$ that are like n-th exterior powers:

  $$\lambda^n(x+y) = \sum_0^n \lambda^i(x)\lambda^{n-i}(y).$$

  For example, Z is a λ-ring with $\lambda^n(x) = \binom{x}{n}$, the binomial coefficients. The notion plays a central rule in the algebraic approach to the Riemann–Roch theorem.
- A totally ordered ring is a ring with a total ordering that is compatible with ring operations.

Some Examples of the Ubiquity of Rings

Many different kinds of mathematical objects can be fruitfully analyzed in terms of some associated ring.

Cohomology Ring of a Topological Space

To any topological space X one can associate its integral cohomology ring

$$H^*(X,\mathbb{Z}) = \bigoplus_{i=0}^{\infty} H^i(X,\mathbb{Z}),$$

a graded ring. There are also homology groups $H_i(X,\mathbb{Z})$ of a space, and indeed these were defined first, as a useful tool for distinguishing between certain pairs of topological spaces, like the spheres and tori, for which the methods of point-set topology are not well-suited. Cohomology groups were later defined in terms of homology groups in a way which is roughly analogous to the dual of a vector space. To know each individual integral homology group is essentially the same as knowing each individual integral cohomology group, because of the universal coefficient theorem. However, the advantage of the cohomology groups is that there is a natural product, which is analogous to the observation that one can multiply pointwise a k-multilinear form and an l-multilinear form to get a $(k + l)$-multilinear form.

The ring structure in cohomology provides the foundation for characteristic classes of fiber bundles, intersection theory on manifolds and algebraic varieties, Schubert calculus and much more.

Burnside Ring of a Group

To any group is associated its Burnside ring which uses a ring to describe the various ways the group can act on a finite set. The Burnside ring's additive group is the free abelian group whose basis are the transitive actions of the group and whose addition is the disjoint union of the action. Expressing an action in terms of the basis is decomposing an action into its transitive constituents. The multiplication is easily expressed in terms of the representation ring: the multiplication in the Burnside ring is formed by writing the tensor product of two permutation modules as a permutation module. The ring structure allows a formal way of subtracting one action from another. Since the Burnside ring is contained as a finite index subring of the representation ring, one can pass easily from one to the other by extending the coefficients from integers to the rational numbers.

Representation Ring of a Group Ring

To any group ring or Hopf algebra is associated its representation ring or "Green ring". The representation ring's additive group is the free abelian group whose basis are the indecomposable modules and whose addition corresponds to the direct sum. Expressing a module in terms of the basis is finding an indecomposable decomposition of the module. The multiplication is the tensor product. When the algebra is semisimple, the representation ring is just the character ring from character theory, which is more or less the Grothendieck group given a ring structure.

Function Field of an Irreducible Algebraic Variety

To any irreducible algebraic variety is associated its function field. The points of an algebraic variety correspond to valuation rings contained in the function field and containing the coordinate ring. The study of algebraic geometry makes heavy use of commutative algebra to study geometric concepts in terms of ring-theoretic properties. Birational geometry studies maps between the subrings of the function field.

Face Ring of a Simplicial Complex

Every simplicial complex has an associated face ring, also called its Stanley–Reisner ring. This ring reflects many of the combinatorial properties of the simplicial complex, so it is of particular interest in algebraic combinatorics. In particular, the algebraic geometry of the Stanley–Reisner ring was used to characterize the numbers of faces in each dimension of simplicial polytopes.

Category Theoretical Description

Every ring can be thought of as a monoid in Ab, the category of abelian groups (thought of as a monoidal category under the tensor product of $\mathbb{Z}$-modules). The monoid action of a ring R on an abelian group is simply an R-module. Essentially, an R-module is a generalization of the notion of a vector space – where rather than a vector space over a field, one has a "vector space over a ring".

Let $(A, +)$ be an abelian group and let End(A) be its endomorphism ring. Note that, essentially, End(A) is the set of all morphisms of A, where if f is in End(A), and g is in End(A), the following rules may be used to compute $f + g$ and $f \cdot g$:

- $(f + g)(x) = f(x) + g(x)$
- $(f \cdot g)(x) = f(g(x))$

where + as in $f(x) + g(x)$ is addition in A, and function composition is denoted from right to left.

Therefore, associated to any abelian group, is a ring. Conversely, given any ring, $(R, +, \cdot)$, $(R, +)$ is an abelian group. Furthermore, for every r in R, right (or left) multiplication by r gives rise to a morphism of $(R, +)$, by right (or left) distributivity. Let $A = (R, +)$. Consider those endomorphisms of A, that "factor through" right (or left) multiplication of R. In other words, let $\mathrm{End}_R(A)$ be the set of all morphisms m of A, having the property that $m(r \cdot x) = r \cdot m(x)$. It was seen that every r in R gives rise to a morphism of A: right multiplication by r. It is in fact true that this association of any element of R, to a morphism of A, as a function from R to $\mathrm{End}_R(A)$, is an isomorphism of rings. In this sense, therefore, any ring can be viewed as the endomorphism ring of some abelian X-group (by X-group, it is meant a group with X being its set of operators). In essence, the most general form of a ring, is the endomorphism group of some abelian X-group.

Any ring can be seen as a preadditive category with a single object. It is therefore natural to consider arbitrary preadditive categories to be generalizations of rings. And indeed, many definitions and theorems originally given for rings can be translated to this more general context. Additive functors between preadditive categories generalize the concept of ring homomorphism, and ideals in additive categories can be defined as sets of morphisms closed under addition and under composition with arbitrary morphisms.

Generalization

Algebraists have defined structures more general than rings by weakening or dropping some of ring axioms.

Rng

A rng is the same as a ring, except that the existence of a multiplicative identity is not assumed.

Nonassociative Ring

A nonassociative ring is an algebraic structure that satisfies all of the ring axioms but the associativity and the existence of a multiplicative identity. A notable example is a Lie algebra. There exists some structure theory for such algebras that generalizes the analogous results for Lie algebras and associative algebras.

Semiring

A semiring is obtained by weakening the assumption that $(R,+)$ is an abelian group to the assumption that $(R,+)$ is a commutative monoid, and adding the axiom that $0 \cdot a = a \cdot 0 = 0$ for all a in R (since it no longer follows from the other axioms).

Example: a tropical semiring.

Other Ring-like Objects

Ring Object in a Category

Let C be a category with finite products. Let pt denote a terminal object of C (an empty product). A ring object in C is an object R equipped with morphisms $R \times R \xrightarrow{a} R$ (addition), $R \times R \xrightarrow{m} R$

(multiplication), $\mathrm{pt}\xrightarrow{0}R$ (additive identity), $R\xrightarrow{i}R$ (additive inverse), and $\mathrm{pt}\xrightarrow{1}R$ (multiplicative identity) satisfying the usual ring axioms. Equivalently, a ring object is an object R equipped with a factorization of its functor of points $h_R = \mathrm{Hom}(-, R) : C^{\mathrm{op}} \to \mathbf{Sets}$ through the category of rings: $C^{\mathrm{op}} \to \mathbf{Rings} \xrightarrow{\text{forgetful}} \mathbf{Sets}$.

Ring Scheme

In algebraic geometry, a ring scheme over a base scheme S is a ring object in the category of S-schemes. One example is the ring scheme W_n over Spec Z, which for any commutative ring A returns the ring $W_n(A)$ of p-isotypic Witt vectors of length n over A.

Ring Spectrum

In algebraic topology, a ring spectrum is a spectrum X together with a multiplication $\mu : X \wedge X \to X$ and a unit map $S \to X$ from the sphere spectrum S, such that the ring axiom diagrams commute up to homotopy. In practice, it is common to define a ring spectrum as a monoid object in a good category of spectra such as the category of symmetric spectra.

Ring Theory

In algebra, ring theory is the study of rings—algebraic structures in which addition and multiplication are defined and have similar properties to those operations defined for the integers. Ring theory studies the structure of rings, their representations, or, in different language, modules, special classes of rings (group rings, division rings, universal enveloping algebras), as well as an array of properties that proved to be of interest both within the theory itself and for its applications, such as homological properties and polynomial identities.

Commutative rings are much better understood than noncommutative ones. Algebraic geometry and algebraic number theory, which provide many natural examples of commutative rings, have driven much of the development of commutative ring theory, which is now, under the name of *commutative algebra*, a major area of modern mathematics. Because these three fields (algebraic geometry, algebraic number theory and commutative algebra) are so intimately connected it is usually difficult and meaningless to decide which field a particular result belongs to. For example, Hilbert's Nullstellensatz is a theorem which is fundamental for algebraic geometry, and is stated and proved in terms of commutative algebra. Similarly, Fermat's last theorem is stated in terms of elementary arithmetic, which is a part of commutative algebra, but its proof involves deep results of both algebraic number theory and algebraic geometry.

Noncommutative rings are quite different in flavour, since more unusual behavior can arise. While the theory has developed in its own right, a fairly recent trend has sought to parallel the commutative development by building the theory of certain classes of noncommutative rings in a geometric fashion as if they were rings of functions on (non-existent) 'noncommutative spaces'. This trend started in the 1980s with the development of noncommutative geometry and with the discovery of quantum groups. It has led to a better understanding of noncommutative rings, especially noncommutative Noetherian rings.

For the definitions of a ring and basic concepts and their properties.

Commutative Rings

A ring is called *commutative* if its multiplication is commutative. Commutative rings resemble familiar number systems, and various definitions for commutative rings are designed to formalize properties of the integers. Commutative rings are also important in algebraic geometry. In commutative ring theory, numbers are often replaced by ideals, and the definition of the prime ideal tries to capture the essence of prime numbers. Integral domains, non-trivial commutative rings where no two non-zero elements multiply to give zero, generalize another property of the integers and serve as the proper realm to study divisibility. Principal ideal domains are integral domains in which every ideal can be generated by a single element, another property shared by the integers. Euclidean domains are integral domains in which the Euclidean algorithm can be carried out. Important examples of commutative rings can be constructed as rings of polynomials and their factor rings. Summary: Euclidean domain => principal ideal domain => unique factorization domain => integral domain => Commutative ring.

Algebraic Geometry

Algebraic geometry is in many ways the mirror image of commutative algebra. This correspondence started with Hilbert's Nullstellensatz that establishes a one-to-one correspondence between the points of an algebraic variety, and the maximal ideals of its coordinate ring. This correspondence has been enlarged and systematized for translating (and proving) most geometrical properties of algebraic varieties into algebraic properties of associated commutative rings. Alexander Grothendieck completed this by introducing schemes, a generalization of algebraic varieties, which may be built from any commutative ring. More precisely, the spectrum of a commutative ring is the space of its prime ideals equipped with Zariski topology, and augmented with a sheaf of rings. These objects are the "affine schemes" (generalization of affine varieties), and a general scheme is then obtained by "gluing together" (by purely algebraic methods) several such affine schemes, in analogy to the way of constructing a manifold by gluing together the charts of an atlas.

Noncommutative Rings

Noncommutative rings resemble rings of matrices in many respects. Following the model of algebraic geometry, attempts have been made recently at defining noncommutative geometry based on noncommutative rings. Noncommutative rings and associative algebras (rings that are also vector spaces) are often studied via their categories of modules. A module over a ring is an abelian group that the ring acts on as a ring of endomorphisms, very much akin to the way fields (integral domains in which every non-zero element is invertible) act on vector spaces. Examples of noncommutative rings are given by rings of square matrices or more generally by rings of endomorphisms of abelian groups or modules, and by monoid rings.

Representation Theory

Representation theory is a branch of mathematics that draws heavily on non-commutative rings. It studies abstract algebraic structures by *representing* their elements as linear transformations of vector spaces, and studies modules over these abstract algebraic structures. In essence, a representation makes an abstract algebraic object more concrete by describing its elements by matrices and the algebraic operations in terms of matrix addition and matrix multiplication, which is non-commutative. The algebraic objects amenable to such a description include groups, associative algebras and Lie algebras. The most prominent of these (and historically the first) is the representation theory of groups, in which elements of a group are represented by invertible matrices in such a way that the group operation is matrix multiplication.

Some Relevant Theorems

General

- Isomorphism theorems for rings.
- Nakayama's lemma.

Structure theorems

- The Artin–Wedderburn theorem determines the structure of semisimple rings.
- The Jacobson density theorem determines the structure of primitive rings.
- Goldie's theorem determines the structure of semiprime Goldie rings.
- The Zariski–Samuel theorem determines the structure of a commutative principal ideal ring.
- The Hopkins–Levitzki theorem gives necessary and sufficient conditions for a Noetherian ring to be an Artinian ring.
- Morita theory consists of theorems determining when two rings have "equivalent" module categories.
- Cartan–Brauer–Hua theorem gives insight on the structure of division rings.
- Wedderburn's little theorem states that finite domains are fields.

Other

- The Skolem–Noether theorem characterizes the automorphisms of simple rings.

Structures and Invariants of Rings

Dimension of a Commutative Ring

The Krull dimension of a commutative ring R is the supremum of the lengths n of all the increasing chains of prime ideals $\mathfrak{p}_0 \subsetneq \mathfrak{p}_1 \subsetneq \cdots \subsetneq \mathfrak{p}_n$.. For example, the polynomial ring $k[t_1,\cdots,t_n]$ over a field k has dimension n. The fundamental theorem in the dimension theory states the following numbers coincide for a noetherian local ring $(R,\mathfrak{m})$:

- The Krull dimension of R.
- The minimum number of the generators of the $\mathfrak{m}$ -primary ideals.

- The dimension of the graded ring $\mathrm{gr}_{\mathfrak{m}}(R) = \oplus_{k\geq 0}\mathfrak{m}^k / \mathfrak{m}^{k+1}$ (equivalently, one plus the degree of its Hilbert polynomial).

A commutative ring *R* is said to be catenary if any pair of prime ideals $\mathfrak{p} \subset \mathfrak{p}'$ can be extended to a chain of prime ideals $\mathfrak{p} = \mathfrak{p}_0 \subsetneq \cdots \subsetneq \mathfrak{p}_n = \mathfrak{p}'$ of same finite length such that there is no prime ideal that is strictly contained in two consecutive terms. Practically all noetherian rings that appear in application are catenary. If $(R, \mathfrak{m})$ is a catenary local integral domain, then, by definition,

$$\dim R = \mathrm{ht}\,\mathfrak{p} + \dim R / \mathfrak{p}$$

where $\mathrm{ht}\,\mathfrak{p} = \dim R_{\mathfrak{p}}$ is the height of $\mathfrak{p}$. It is a deep theorem of Ratliff that the converse is also true.

If *R* is an integral domain that is a finitely generated *k*-algebra, then its dimension is the transcendence degree of its field of fractions over *k*. If *S* is an integral extension of a commutative ring *R*, then *S* and *R* have the same dimension.

Closely related concepts are those of depth and global dimension. In general, if *R* is a noetherian local ring, then the depth of *R* is less than or equal to the dimension of *R*. When the equality holds, *R* is called a Cohen–Macaulay ring. A regular local ring is an example of a Cohen–Macaulay ring. It is a theorem of Serre that *R* is a regular local ring if and only if it has finite global dimension and in that case the global dimension is the Krull dimension of *R*. The significance of this is that a global dimension is a homological notion.

Morita Equivalence

Two rings *R*, *S* are said to be Morita equivalent if the category of left modules over *R* is equivalent to the category of left modules over *S*. In fact, two commutative rings which are Morita equivalent must be isomorphic, so the notion does not add anything new to the category of commutative rings. However, commutative rings can be Morita equivalent to noncommutative rings, so Morita equivalence is coarser than isomorphism. Morita equivalence is especially important in algebraic topology and functional analysis.

Finitely Generated Projective Module Over a Ring and Picard Group

Let *R* be a commutative ring and $\mathbf{P}(R)$ the set of isomorphism classes of finitely generated projective modules over *R*; let also $\mathbf{P}_n(R)$ subsets consisting of those with constant rank *n*. (The rank of a module *M* is the continuous function $\mathrm{Spec}\, R \to \mathbb{Z}, \mathfrak{p} \mapsto \dim M \otimes_R k(\mathfrak{p})$.) $\mathbf{P}_1(R)$ is usually denoted by Pic(*R*). It is an abelian group called the Picard group of *R*. If *R* is an integral domain with the field of fractions *F* of *R*, then there is an exact sequence of groups:

$$1 \to R^* \to F^* \xrightarrow{f \mapsto fR} \mathrm{Cart}(R) \to \mathrm{Pic}(R) \to 1$$

where $\mathrm{Cart}(R)$ is the set of fractional ideals of *R*. If *R* is a regular domain (i.e., regular at any prime ideal), then Pic(R) is precisely the divisor class group of *R*.

For example, if R is a principal ideal domain, then Pic(R) vanishes. In algebraic number theory, R will be taken to be the ring of integers, which is Dedekind and thus regular. It follows that Pic(R) is a finite group (finiteness of class number) that measures the deviation of the ring of integers from being a PID.

One can also consider the group completion of **P**(R); this results in a commutative ring $K_0(R)$. Note that $K_0(R) = K_0(S)$ if two commutative rings R, S are Morita equivalent.

Structure of Noncommutative Rings

The structure of a noncommutative ring is more complicated than that of a commutative ring. For example, there exist simple rings, containing no non-trivial proper (two-sided) ideals, which contain non-trivial proper left or right ideals. Various invariants exist for commutative rings, whereas invariants of noncommutative rings are difficult to find. As an example, the nilradical of a ring, the set of all nilpotent elements, need not be an ideal unless the ring is commutative. Specifically, the set of all nilpotent elements in the ring of all n x n matrices over a division ring never forms an ideal, irrespective of the division ring chosen. There are, however, analogues of the nilradical defined for noncommutative rings, that coincide with the nilradical when commutativity is assumed.

The concept of the Jacobson radical of a ring; that is, the intersection of all right/left annihilators of simple right/left modules over a ring, is one example. The fact that the Jacobson radical can be viewed as the intersection of all maximal right/left ideals in the ring, shows how the internal structure of the ring is reflected by its modules. It is also a fact that the intersection of all maximal right ideals in a ring is the same as the intersection of all maximal left ideals in the ring, in the context of all rings; whether commutative or noncommutative.

Noncommutative rings serve as an active area of research due to their ubiquity in mathematics. For instance, the ring of n-by-n matrices over a field is noncommutative despite its natural occurrence in geometry, physics and many parts of mathematics. More generally, endomorphism rings of abelian groups are rarely commutative, the simplest example being the endomorphism ring of the Klein four-group.

One of the best known noncommutative rings is the division ring of quaternions.

Applications

The Coordinate Ring of an Algebraic Variety

If X is an affine algebraic variety, then the set of all regular functions on X forms a ring called the coordinate ring of X. For a projective variety, there is an analogous ring called the homogeneous coordinate ring. Those rings are essentially the same things as varieties: they correspond in essentially a unique way. This may be seen via either Hilbert's Nullstellensatz or scheme-theoretic constructions (i.e., Spec and Proj).

Ring of Invariants

A basic (and perhaps the most fundamental) question in the classical invariant theory is to find and study polynomials in the polynomial ring $k[V]$ that are invariant under the action of a finite group (or more generally reductive) G on V. The main example is the ring of symmetric polynomials: symmetric polynomials are polynomials that are invariant under permutation of variable. The fundamental theorem of symmetric polynomials states that this ring is $R[\sigma_1,\ldots,\sigma_n]$ where σ_i are elementary symmetric polynomials.

Lattice

A lattice is an abstract structure studied in the mathematical subdisciplines of order theory and abstract algebra. It consists of a partially ordered set in which every two elements have a unique supremum (also called a least upper bound or join) and a unique infimum (also called a greatest lower bound or meet). An example is given by the natural numbers, partially ordered by divisibility, for which the unique supremum is the least common multiple and the unique infimum is the greatest common divisor.

Lattices can also be characterized as algebraic structures satisfying certain axiomatic identities. Since the two definitions are equivalent, lattice theory draws on both order theory and universal algebra. Semilattices include lattices, which in turn include Heyting and Boolean algebras. These "lattice-like" structures all admit order-theoretic as well as algebraic descriptions.

Lattices as Partially Ordered Sets

If $(L, \leq)$ is a partially ordered set (poset), and $S \subseteq L$ is an arbitrary subset, then an element $u \in L$ is said to be an upper bound of S if $s \leq u$ for each $s \in S$. A set may have many upper bounds, or none at all. An upper bound u of S is said to be its least upper bound, or join, or supremum, if $u \leq x$ for each upper bound x of S. A set need not have a least upper bound, but it cannot have more than one. Dually, $l \in L$ is said to be a lower bound of S if $l \leq s$ for each $s \in S$. A lower bound l of S is said to be its greatest lower bound, or meet, or infimum, if $x \leq l$ for each lower bound x of S. A set may have many lower bounds, or none at all, but can have at most one greatest lower bound.

A partially ordered set $(L, \leq)$ is called a join-semilattice if each two-element subset $\{a, b\} \subseteq L$ has a join (i.e. least upper bound), and is called a meet-semilattice if each two-element subset has a meet (i.e. greatest lower bound), denoted by $a \vee b$ and $a \wedge b$ respectively. $(L, \leq)$ is called a lattice if it is both a join- and a meet-semilattice. This definition makes $\vee$ and $\wedge$ binary operations. Both operations are monotone with respect to the order: $a_1 \leq a_2$ and $b_1 \leq b_2$ implies that $a_1 \vee b_1 \leq a_2 \vee b_2$ and $a_1 \wedge b_1 \leq a_2 \wedge b_2$.

It follows by an induction argument that every non-empty finite subset of a lattice has a least upper bound and a greatest lower bound. With additional assumptions, further conclusions may be possible. That topic also discusses how one may rephrase the above definition in terms of the existence of suitable Galois connections between related partially ordered sets — an approach of special interest for the category theoretic approach to lattices, and for formal concept analysis.

A bounded lattice is a lattice that additionally has a greatest element 1 and a least element 0, which satisfy

$0 \le x \le 1$ for every x in L.

The greatest and least element is also called the maximum and minimum, or the top and bottom element, and denoted by $\top$ and $\bot$, respectively. Every lattice can be converted into a bounded lattice by adding an artificial greatest and least element, and every non-empty finite lattice is bounded, by taking the join (resp., meet) of all elements, denoted by $\bigvee L = a_1 \vee \cdots \vee a_n$ $(\bigwedge L = a_1 \wedge \cdots \wedge a_n)$ where $L = \{a_1, \ldots, a_n\}$.

A partially ordered set is a bounded lattice if and only if every finite set of elements (including the empty set) has a join and a meet. For every element x of a poset it is trivially true (it is a vacuous truth) that $\forall a \in \varnothing : x \le a$ and $\forall a \in \varnothing : a \le x$, and therefore every element of a poset is both an upper bound and a lower bound of the empty set. This implies that the join of an empty set is the least element $\bigvee \varnothing = 0$, and the meet of the empty set is the greatest element $\bigwedge \varnothing = 1$. This is consistent with the associativity and commutativity of meet and join: the join of a union of finite sets is equal to the join of the joins of the sets, and dually, the meet of a union of finite sets is equal to the meet of the meets of the sets, i.e., for finite subsets A and B of a poset L,

$$\bigvee(A \cup B) = \left(\bigvee A\right) \vee \left(\bigvee B\right)$$

and

$$\bigwedge(A \cup B) = \left(\bigwedge A\right) \wedge \left(\bigwedge B\right)$$

hold. Taking B to be the empty set,

$$\bigvee(A \cup \varnothing) = \left(\bigvee A\right) \vee \left(\bigvee \varnothing\right) = \left(\bigvee A\right) \vee 0 = \bigvee A$$

and

$$\bigwedge(A \cup \varnothing) = \left(\bigwedge A\right) \wedge \left(\bigwedge \varnothing\right) = \left(\bigwedge A\right) \wedge 1 = \bigwedge A$$

which is consistent with the fact that $A \cup \varnothing = A$.

A lattice element y is said to cover another element x, if $y > x$, but there does not exist a z such that $y > z > x$. Here, $y > x$ means $x \le y$ and $x \ne y$.

A lattice $(L, \le)$ is called graded, sometimes ranked, if it can be equipped with a rank function r from L to $\mathbb{N}$, sometimes to $\mathbb{Z}$, compatible with the ordering (so $r(x) < r(y)$ whenever $x < y$) such that whenever y covers x, then $r(y) = r(x) + 1$. The value of the rank function for a lattice element is called its rank.

Given a subset of a lattice, $H \subset L$, meet and join restrict to partial functions – they are undefined if their value is not in the subset H. The resulting structure on H is called a partial lattice. In addition to this extrinsic definition as a subset of some other algebraic structure (a lattice), a partial lattice can also be intrinsically defined as a set with two partial binary operations satisfying certain axioms.

Lattices as Algebraic Structures

General Lattice

An algebraic structure $(L, \vee, \wedge)$, consisting of a set L and two binary operations $\vee$, and $\wedge$, on L is a lattice if the following axiomatic identities hold for all elements a, b, c of L.

Commutative laws	Associative laws	Absorption laws
$a \vee b = b \vee a,$	$a \vee (b \vee c) = (a \vee b) \vee c,$	$a \vee (a \wedge b) = a,$
$a \wedge b = b \wedge a.$	$a \wedge (b \wedge c) = (a \wedge b) \wedge c.$	$a \wedge (a \vee b) = a.$

The following two identities are also usually regarded as axioms, even though they follow from the two absorption laws taken together.

Idempotent laws

$$a \vee a = a,$$

$$a \wedge a = a.$$

These axioms assert that both $(L, \vee)$ and $(L, \wedge)$ are semilattices. The absorption laws, the only axioms above in which both meet and join appear, distinguish a lattice from an arbitrary pair of semilattices and assure that the two semilattices interact appropriately. In particular, each semilattice is the dual of the other.

Bounded Lattice

A bounded lattice is an algebraic structure of the form $(L, \vee, \wedge, 0, 1)$ such that $(L, \vee, \wedge)$ is a lattice, 0 (the lattice's bottom) is the identity element for the join operation $\vee$, and 1 (the lattice's top) is the identity element for the meet operation $\wedge$.

Identity laws

$$a \vee 0 = a,$$

$$a \wedge 1 = a.$$

Connection to other Algebraic Structures

Lattices have some connections to the family of group-like algebraic structures. Because meet and join both commute and associate, a lattice can be viewed as consisting of two commutative semigroups having the same domain. For a bounded lattice, these semigroups are in fact commutative monoids. The absorption law is the only defining identity that is peculiar to lattice theory.

By commutativity and associativity one can think of join and meet as binary operations that are defined on non-empty finite sets, rather than on elements. In a bounded lattice the empty join and the empty meet can also be defined (as 0 and 1, respectively). This makes bounded lattices somewhat more natural than general lattices, and many authors require all lattices to be bounded.

The algebraic interpretation of lattices plays an essential role in universal algebra.

Connection between the Two Definitions

An order-theoretic lattice gives rise to the two binary operations ∨ and ∧. Since the commutative, associative and absorption laws can easily be verified for these operations, they make $(L, \vee, \wedge)$ into a lattice in the algebraic sense.

The converse is also true. Given an algebraically defined lattice $(L, \vee, \wedge)$, one can define a partial order ≤ on L by setting

$$a \leq b \text{ if } a = a \wedge b, \text{ or}$$

$$a \leq b \text{ if } b = a \vee b,$$

for all elements a and b from L. The laws of absorption ensure that both definitions are equivalent:

$a = a \wedge b$ implies $b = b \vee (b \wedge a) = (a \wedge b) \vee b = a \vee b$

and dually for the other direction.

One can now check that the relation ≤ introduced in this way defines a partial ordering within which binary meets and joins are given through the original operations ∨ and ∧.

Since the two definitions of a lattice are equivalent, one may freely invoke aspects of either definition in any way that suits the purpose at hand.

Examples

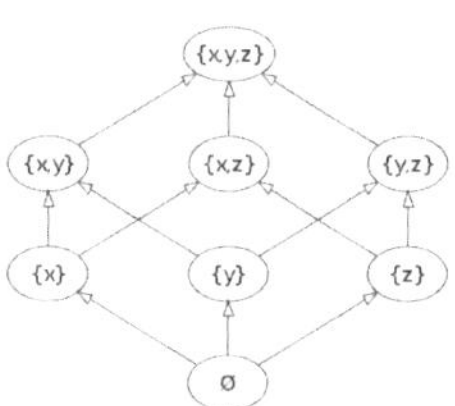

Pic.1: Lattice of subsets of {x, y, z}, ordered by "*is subset of*". The name "lattice" is suggested by the form of the Hasse diagram depicting it.

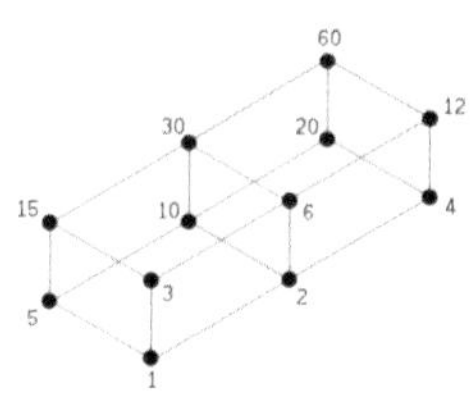

Pic.2: Lattice of integer divisors of 60, ordered by "*divides*".

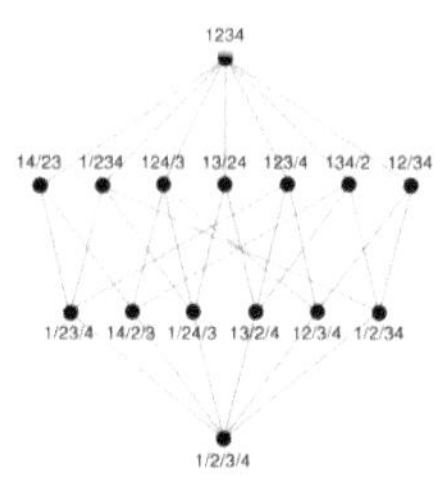

Pic.3: Lattice of partitions of {1, 2, 3, 4}, ordered by "*refines*".

Pic.4: Lattice of positive integers, ordered by <.

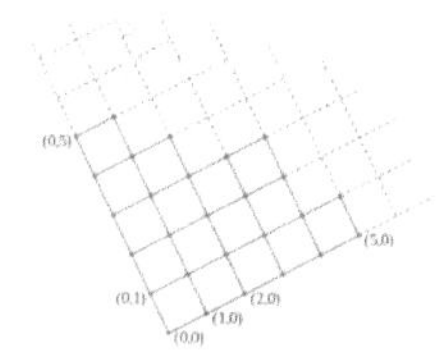
Pic.5: Lattice of nonnegative integer pairs, ordered componentwise.

- For any set A, the collection of all subsets of A (called the power set of A) can be ordered via subset inclusion to obtain a lattice bounded by A itself and the null set. Set intersection and union interpret meet and join, respectively (see pic.1).
- For any set A, the collection of all finite subsets of A, ordered by inclusion, is also a lattice, and will be bounded if and only if A is finite.
- For any set A, the collection of all partitions of A, ordered by refinement, is a lattice.

- The positive integers in their usual order form a lattice, under the operations of "min" and "max". 1 is bottom; there is no top.
- The Cartesian square of the natural numbers, ordered so that $(a, b) \le (c, d)$ if $a \le c$ and $b \le d$. The pair (0, 0) is the bottom element; there is no top.
- The natural numbers also form a lattice under the operations of taking the greatest common divisor and least common multiple, with divisibility as the order relation: $a \le b$ if a divides b. 1 is bottom; 0 is top. Pic.2 shows a finite sublattice.
- Every complete lattice is a (rather specific) bounded lattice. This class gives rise to a broad range of practical examples.
- The set of compact elements of an arithmetic complete lattice is a lattice with a least element, where the lattice operations are given by restricting the respective operations of the arithmetic lattice. This is the specific property which distinguishes arithmetic lattices from algebraic lattices, for which the compacts do only form a join-semilattice. Both of these classes of complete lattices are studied in domain theory.

Further examples of lattices are given for each of the additional properties discussed below.

Counter-examples

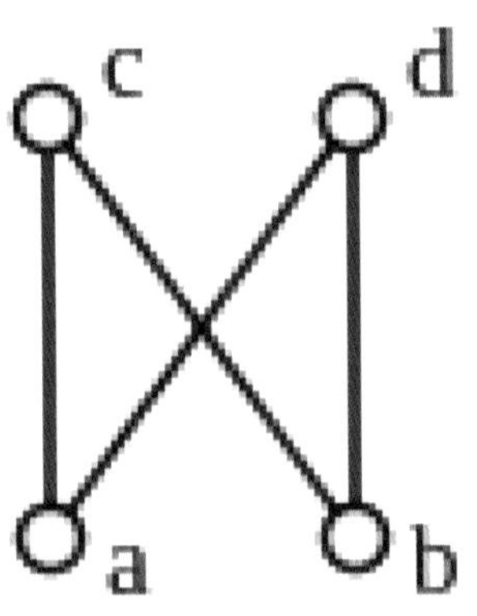

Pic.6: Non-lattice poset: c and d have no common upper bound.

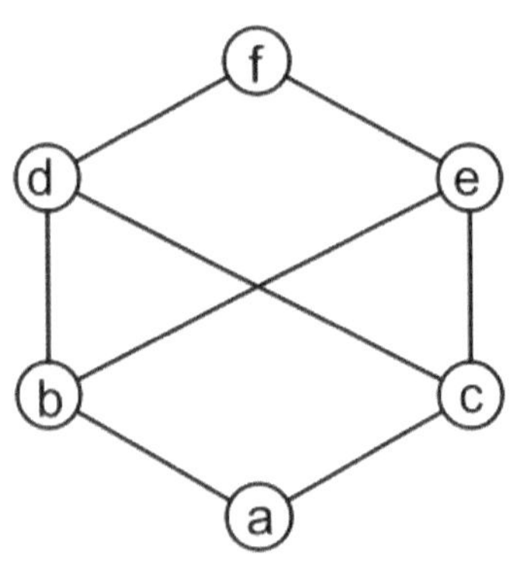

Pic.7: Non-lattice poset: b and c have common upper bounds d, e, and f, but none of them is the least upper bound.

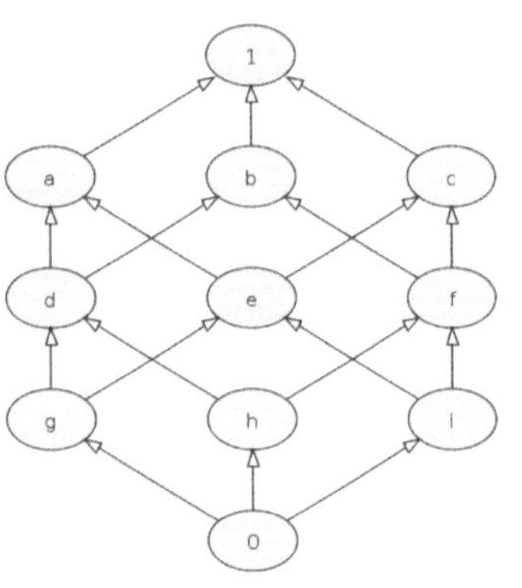

Pic.8: Non-lattice poset: a and b have common lower bounds 0, d, g, h, and i, but none of them is the greatest lower bound.

Most partial ordered sets are not lattices, including the following.

- A discrete poset, meaning a poset such that $x \le y$ implies $x = y$, is a lattice if and only if it has at most one element. In particular the two-element discrete poset is not a lattice.
- Although the set {1, 2, 3, 6} partially ordered by divisibility is a lattice, the set {1, 2, 3} so ordered is not a lattice because the pair 2, 3 lacks a join, and it lacks a meet in {2, 3, 6}.
- The set {1, 2, 3, 12, 18, 36} partially ordered by divisibility is not a lattice. Every pair of elements has an upper bound and a lower bound, but the pair 2, 3 has three upper bounds, namely 12, 18, and 36, none of which is the least of those three under divisibility (12 and 18 do not divide

each other). Likewise the pair 12, 18 has three lower bounds, namely 1, 2, and 3, none of which is the greatest of those three under divisibility (2 and 3 do not divide each other).

Morphisms of Lattices

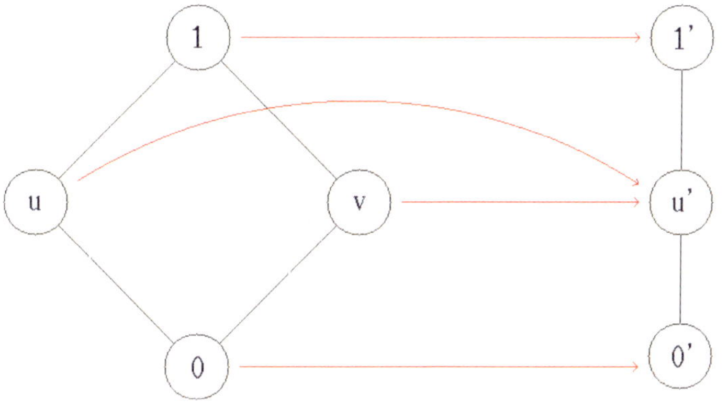

Pic.9: Monotonic map f between lattices that preserves neither joins nor meets, since $f(u) \vee f(v) = u' \vee u' = u' \neq 1' = f(1) = f(u \vee v)$ and $f(u) \wedge f(v) = u' \wedge u' = u' \neq 0' = f(0) = f(u \wedge v)$.

The appropriate notion of a morphism between two lattices flows easily from the above algebraic definition. Given two lattices $(L, \vee_L, \wedge_L)$ and $(M, \vee_M, \wedge_M)$, a lattice homomorphism from L to M is a function $f : L \rightarrow M$ such that for all $a, b \in L$:

$$f(a \vee_L b) = f(a) \vee_M f(b), \text{ and}$$

$$f(a \wedge_L b) = f(a) \wedge_M f(b).$$

Thus f is a homomorphism of the two underlying semilattices. When lattices with more structure are considered, the morphisms should "respect" the extra structure, too. In particular, a bounded-lattice homomorphism (usually called just "lattice homomorphism") f between two bounded lattices L and M should also have the following property:

$$f(0_L) = 0_M, \text{ and}$$

$$f(1_L) = 1_M.$$

In the order-theoretic formulation, these conditions just state that a homomorphism of lattices is a function preserving binary meets and joins. For bounded lattices, preservation of least and greatest elements is just preservation of join and meet of the empty set.

Any homomorphism of lattices is necessarily monotone with respect to the associated ordering relation. The converse is not true: monotonicity by no means implies the required preservation of meets and joins (see pic.9), although an order-preserving bijection is a homomorphism if its inverse is also order-preserving.

Given the standard definition of isomorphisms as invertible morphisms, a *lattice isomorphism* is just a bijective lattice homomorphism. Similarly, a *lattice endomorphism* is a lattice homomorphism from a lattice to itself, and a *lattice automorphism* is a bijective lattice endomorphism. Lattices and their homomorphisms form a category.

Sublattices

A *sublattice* of a lattice L is a nonempty subset of L that is a lattice with the same meet and join operations as L. That is, if L is a lattice and $M \neq \varnothing$ is a subset of L such that for every pair of elements a, b in M both $a \wedge b$ and $a \vee b$ are in M, then M is a sublattice of L.

A sublattice M of a lattice L is a *convex sublattice* of L, if $x \leq z \leq y$ and x, y in M implies that z belongs to M, for all elements x, y, z in L.

Properties of Lattices

We now introduce a number of important properties that lead to interesting special classes of lattices. One, boundedness, has already been discussed.

Completeness

A poset is called a complete lattice if *all* its subsets have both a join and a meet. In particular, every complete lattice is a bounded lattice. While bounded lattice homomorphisms in general preserve only finite joins and meets, complete lattice homomorphisms are required to preserve arbitrary joins and meets.

Every poset that is a complete semilattice is also a complete lattice. Related to this result is the interesting phenomenon that there are various competing notions of homomorphism for this class of posets, depending on whether they are seen as complete lattices, complete join-semilattices, complete meet-semilattices, or as join-complete or meet-complete lattices.

Note that "partial lattice" is not the opposite of "complete lattice" – rather, "partial lattice", "lattice", and "complete lattice" are increasingly restrictive definitions.

Conditional Completeness

A conditionally complete lattice is a lattice in which every *nonempty* subset *that has an upper bound* has a join (i.e., a least upper bound). Such lattices provide the most direct generalization of the completeness axiom of the real numbers. A conditionally complete lattice is either a complete lattice, or a complete lattice without its maximum element 1, its minimum element 0, or both.

Distributivity

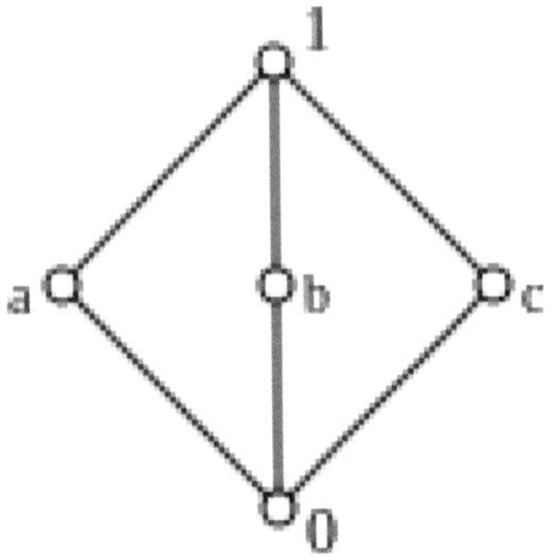

Pic.10: Smallest non-distributive (but modular) lattice M_3.

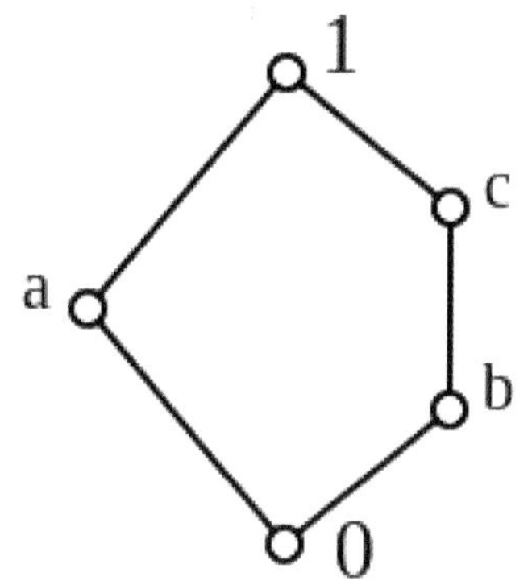

Pic.11: Smallest non-modular (and hence non-distributive) lattice N_5.
The labelled elements violate the distributivity equation $c \wedge (a \vee b) = (c \wedge a) \vee (c \wedge b)$,
but satisfy its dual $c \vee (a \wedge b) = (c \vee a) \wedge (c \vee b)$.

Since lattices come with two binary operations, it is natural to ask whether one of them distributes over the other, i.e. whether one or the other of the following dual laws holds for every three elements a, b, c of L:

Distributivity of $\vee$ over $\wedge$

$$a \vee (b \wedge c) = (a \vee b) \wedge (a \vee c).$$

Distributivity of $\wedge$ over $\vee$

$$a \wedge (b \vee c) = (a \wedge b) \vee (a \wedge c).$$

A lattice that satisfies the first or, equivalently (as it turns out), the second axiom, is called a distributive lattice. The only non-distributive lattices with fewer than 6 elements are called M_3 and N_5; they are shown in picture 10 and 11, respectively. A lattice is distributive if and only if it doesn't have a sublattice isomorphic to M_3 or N_5. Each distributive lattice is isomorphic to a lattice of sets (with union and intersection as join and meet, respectively).

For an overview of stronger notions of distributivity which are appropriate for complete lattices and which are used to define more special classes of lattices such as frames and completely distributive lattices.

Modularity

For some applications the distributivity condition is too strong, and the following weaker property is often useful. A lattice $(L, \vee, \wedge)$ is modular if, for all elements a, b, c of L, the following identity holds.

Modular Identity

$$(a \wedge c) \vee (b \wedge c) = [(a \wedge c) \vee b] \wedge c.$$

This condition is equivalent to the following axiom.

Modular Law

$$a \leq c \text{ implies } a \vee (b \wedge c) = (a \vee b) \wedge c.$$

A lattice is modular if and only if it doesn't have a sublattice isomorphic to N_5 (shown in pic.11). Besides distributive lattices, examples of modular lattices are the lattice of two-sided ideals of a ring, the lattice of submodules of a module, and the lattice of normal subgroups of a group. The set of first-order terms with the ordering "*is more specific than*" is a non-modular lattice used in automated reasoning.

Semimodularity

A finite lattice is modular if and only if it is both upper and lower semimodular. For a graded lattice, (upper) semimodularity is equivalent to the following condition on the rank function r:

$$r(x) + r(y) \geq r(x \wedge y) + r(x \vee y).$$

Another equivalent (for graded lattices) condition is Birkhoff's condition:

for each x and y in L, if x and y both cover $x \wedge y$, then $x \vee y$ covers both x and y.

A lattice is called lower semimodular if its dual is semimodular. For finite lattices this means that the previous conditions hold with $\vee$ and $\wedge$ exchanged, "covers" exchanged with "is covered by", and inequalities reversed.

Continuity and Algebraicity

In domain theory, it is natural to seek to approximate the elements in a partial order by "much simpler" elements. This leads to the class of continuous posets, consisting of posets where every element can be obtained as the supremum of a directed set of elements that are way-below the element. If one can additionally restrict these to the compact elements of a poset for obtaining these directed sets, then the poset is even algebraic. Both concepts can be applied to lattices as follows:

- A continuous lattice is a complete lattice that is continuous as a poset.
- An algebraic lattice is a complete lattice that is algebraic as a poset.

Both of these classes have interesting properties. For example, continuous lattices can be characterized as algebraic structures (with infinitary operations) satisfying certain identities. While such a characterization is not known for algebraic lattices, they can be described "syntactically" via Scott information systems.

Complements and Pseudo-complements

Let L be a bounded lattice with greatest element 1 and least element 0. Two elements x and y of L are complements of each other if and only if:

$$x \vee y = 1 \text{ and } x \wedge y = 0.$$

In general, some elements of a bounded lattice might not have a complement, and others might have more than one complement. For example, the set {0, ½, 1} with its usual ordering is a bounded lattice, and ½ does not have a complement. In the bounded lattice N_5, the element a has two complements, viz. b and c (see Pic.11). A bounded lattice for which every element has a complement is called a complemented lattice.

A complemented lattice that is also distributive is a Boolean algebra. For a distributive lattice, the complement of x, when it exists, is unique.

In the case the complement is unique, we write $\neg x = y$ and equivalently, $\neg y = x$. The corresponding unary operation over L, called complementation, introduces an analogue of logical negation into lattice theory.

Heyting algebras are an example of distributive lattices where some members might be lacking complements. Every element x of a Heyting algebra has, on the other hand, a *pseudo-complement*, also denoted $\neg x$. The pseudo-complement is the greatest element y such that $x \wedge y = 0$. If the pseudo-complement of every element of a Heyting algebra is in fact a complement, then the Heyting algebra is in fact a Boolean algebra.

Jordan–Dedekind Chain Condition

A chain from x_0 to x_n is a set $\{x_0, x_1, \ldots, x_n\}$ where $x_0 < x_1 < x_2 < \ldots < x_n$.. The length of this chain is n, or one less than its number of elements. A chain is maximal if x_i covers x_{i-1} for all $1 \le i \le n$.

If for any pair, x and y, where $x < y$, all maximal chains from x to y have the same length, then the lattice is said to satisfy the Jordan–Dedekind chain condition.

Free Lattices

Any set X may be used to generate the free semilattice FX. The free semilattice is defined to consist of all of the finite subsets of X, with the semilattice operation given by ordinary set union. The free semilattice has the universal property. For the free lattice over a set X, Whitman gave a construction based on polynomials over X's members.

Important Lattice-theoretic Notions

We now define some order-theoretic notions of importance to lattice theory. In the following, let x be an element of some lattice L. If L has a bottom element 0, $x \neq 0$ is sometimes required. x is called:

- Join irreducible if $x = a \vee b$ implies $x = a$ or $x = b$ for all a, b in L. When the first condition is generalized to arbitrary joins $\bigvee_{i \in I} a_i$, x is called completely join irreducible (or $\vee$-irreducible). The dual notion is meet irreducibility ($\wedge$-irreducible). For example, in pic.2, the elements 2, 3, 4, and 5 are join irreducible, while 12, 15, 20, and 30 are meet irreducible. In the lattice of real numbers with the usual order, each element is join irreducible, but none is completely join irreducible.

- Join prime if $x \le a \vee b$ implies $x \le a$ or $x \le b$. This too can be generalized to obtain the notion completely join prime. The dual notion is meet prime. Every join-prime element is also join irreducible, and every meet-prime element is also meet irreducible. The converse holds if L is distributive.

Let L have a bottom element 0. An element x of L is an atom if $0 < x$ and there exists no element y of L such that $0 < y < x$. Then L is called:

- Atomic if for every nonzero element x of L, there exists an atom a of L such that $a \leq x$;
- Atomistic if every element of L is a supremum of atoms. That is, for all a, b in L such that $a \leq b$, there exists an atom x of L such that $x \leq a$ and $x \nleq b$.

The notions of ideals and the dual notion of filters refer to particular kinds of subsets of a partially ordered set, and are therefore important for lattice theory. Details can be found in the respective entries.

Module

In mathematics, a module is one of the fundamental algebraic structures used in abstract algebra. A module over a ring is a generalization of the notion of vector space over a field, wherein the corresponding scalars are the elements of an arbitrary given ring (with identity) and a multiplication (on the left and/or on the right) is defined between elements of the ring and elements of the module.

Thus, a module, like a vector space, is an additive abelian group; a product is defined between elements of the ring and elements of the module that is distributive over the addition operation of each parameter and is compatible with the ring multiplication.

Modules are very closely related to the representation theory of groups. They are also one of the central notions of commutative algebra and homological algebra, and are used widely in algebraic geometry and algebraic topology.

Introduction and Definition

Motivation

In a vector space, the set of scalars is a field and acts on the vectors by scalar multiplication, subject to certain axioms such as the distributive law. In a module, the scalars need only be a ring, so the module concept represents a significant generalization. In commutative algebra, both ideals and quotient rings are modules, so that many arguments about ideals or quotient rings can be combined into a single argument about modules. In non-commutative algebra the distinction between left ideals, ideals, and modules becomes more pronounced, though some ring-theoretic conditions can be expressed either about left ideals or left modules.

Much of the theory of modules consists of extending as many of the desirable properties of vector spaces as possible to the realm of modules over a "well-behaved" ring, such as a principal ideal domain. However, modules can be quite a bit more complicated than vector spaces; for instance, not all modules have a basis, and even those that do, free modules, need not have a unique rank if the underlying ring does not satisfy the invariant basis number condition, unlike vector spaces, which always have a (possibly infinite) basis whose cardinality is then unique. (These last two assertions require the axiom of choice in general, but not in the case of finite-dimensional spaces, or certain well-behaved infinite-dimensional spaces such as L^p spaces.)

Formal Definition

Suppose that R is a ring and 1_R is its multiplicative identity. A left R-module M consists of an abelian group $(M, +)$ and an operation $\cdot : R \times M \to M$ such that for all r, s in R and x, y in M, we have:

1. $r \cdot (x + y) = r \cdot x + r \cdot y$
2. $(r + s) \cdot x = r \cdot x + s \cdot x$
3. $(rs) \cdot x = r \cdot (s \cdot x)$
4. $1_R \cdot x = x.$

The operation of the ring on M is called *scalar multiplication*, and is usually written by juxtaposition, i.e. as rx for r in R and x in M, though here it is denoted as $r \cdot x$ to distinguish it from the ring multiplication operation, denoted here by juxtaposition. The notation ${}_RM$ indicates a left R-module M. A right R-module M or M_R is defined similarly, except that the ring acts on the right; i.e., scalar multiplication takes the form $\cdot : M \times R \to M$, and the above axioms are written with scalars r and s on the right of x and y.

Authors who do not require rings to be unital omit condition 4 above in the definition of an R-module, and so would call the structures defined above "unital left R-modules".

If one writes the scalar action as f_r so that $f_r(x) = r \cdot x$, and f for the map that takes each r to its corresponding map f_r, then the first axiom states that every f_r is a group endomorphism of M, and the other three axioms assert that the map $f : R \to \text{End}(M)$ given by $r \mapsto f_r$ is a ring homomorphism from R to the endomorphism ring End(M). Thus a module is a ring action on an abelian group (cf. group action. Also consider monoid action of multiplicative structure of R). In this sense, module theory generalizes representation theory, which deals with group actions on vector spaces, or equivalently group ring actions.

A bimodule is a module that is a left module and a right module such that the two multiplications are compatible.

If R is commutative, then left R-modules are the same as right R-modules and are simply called R-modules.

Examples

- If K is a field, then the concepts "K-vector space" (a vector space over K) and K-module are identical.
- If K is a field, and $K[x]$ a univariate polynomial ring, then a $K[x]$-module M is a K-module with an additional action of x on M that commutes with the action of K on M. In other words, a $K[x]$-module is a K-vector space M combined with a linear map from M to M. Applying the Structure theorem for finitely generated modules over a principal ideal domain to this example shows the existence of the rational and Jordan canonical forms.

- The concept of a Z-module agrees with the notion of an abelian group. That is, every abelian group is a module over the ring of integers Z in a unique way. For $n > 0$, let $n \cdot x = x + x + ... + x$ (n summands), $0 \cdot x = 0$, and $(-n) \cdot x = -(n \cdot x)$. Such a module need not have a basis—groups containing torsion elements do not. (For example, in the group of integers modulo 3, one cannot find even one element which satisfies the definition of a linearly independent set since when an integer such as 3 or 6 multiplies an element the result is 0. However, if a finite field is considered as a module over the same finite field taken as a ring, it is a vector space and does have a basis.)

- The decimal fractions (including negative ones) form a module over the integers. Only singletons are linearly independent sets, but there is no singleton that can serve as a basis, so the module has no basis and no rank.

- If R is any ring and n a natural number, then the cartesian product R^n is both a left and a right module over R if we use the component-wise operations. Hence when $n = 1$, R is an R-module, where the scalar multiplication is just ring multiplication. The case $n = 0$ yields the trivial R-module $\{0\}$ consisting only of its identity element. Modules of this type are called free and if R has invariant basis number (e.g. any commutative ring or field) the number n is then the rank of the free module.

- If R is any ring, $M_n(R)$ is the ring of $n \times n$ matrices over R, let M be an $M_n(R)$-module, and let e_i be the $n \times n$ matrix that has 1 in the (i, i)-entry, and 0 elsewhere. Then e_iM is an R-module, since $re_im = e_irm \in e_iM$. So M breaks up as the direct sum of R-modules, $M = e_1M \oplus ... \oplus e_nM$. Conversely, given an R-module M_0, then $M_0^{\oplus n}$ is an $M_n(R)$-module. In fact, the category of R-module and the category of $M_n(R)$-module are equivalent. The special case is that the module M is just R as a module over itself, then R^n is an $M_n(R)$-module.

- If S is a nonempty set, M is a left R-module, and M^S is the collection of all functions $f : S \rightarrow M$, then with addition and scalar multiplication in M^S defined by $(f + g)(s) = f(s) + g(s)$ and $(rf)(s) = rf(s)$, M^S is a left R-module. The right R-module case is analogous. In particular, if R is commutative then the collection of *R-module homomorphisms* $h : M \rightarrow N$ is an R-module (and in fact a *submodule* of N^M).

- If X is a smooth manifold, then the smooth functions from X to the real numbers form a ring $C^\infty(X)$. The set of all smooth vector fields defined on X form a module over $C^\infty(X)$, and so do the tensor fields and the differential forms on X. More generally, the sections of any vector bundle form a projective module over $C^\infty(X)$, and by Swan's theorem, every projective module is isomorphic to the module of sections of some bundle; the category of $C^\infty(X)$-modules and the category of vector bundles over X are equivalent.

- If R is any ring and I is any left ideal in R, then I is a left module over R. Analogously of course, right ideals are right modules.

- If R is a ring, we can define the ring R^{op} which has the same underlying set and the same addition operation, but the opposite multiplication: if $ab = c$ in R, then $ba = c$ in R^{op}. Any *left* R-module M can then be seen to be a *right* module over R^{op}, and any right module over R can be considered a left module over R^{op}.

- There are modules of a Lie algebra as well.

Submodules and Homomorphisms

Suppose M is a left R-module and N is a subgroup of M. Then N is a submodule (or R-submodule, to be more explicit) if, for any n in N and any r in R, the product $r \cdot n$ is in N (or $n \cdot r$ for a right module).

The set of submodules of a given module M, together with the two binary operations + and ∩, forms a lattice which satisfies the modular law: Given submodules U, N_1, N_2 of M such that $N_1 \subset N_2$, then the following two submodules are equal: $(N_1 + U) \cap N_2 = N_1 + (U \cap N_2)$.

If M and N are left R-modules, then a map $f: M \rightarrow N$ is a homomorphism of ***R***-modules if, for any m, n in M and r, s in R,

$$f(r \cdot m + s \cdot n) = r \cdot f(m) + s \cdot f(n)$$

This, like any homomorphism of mathematical objects, is just a mapping which preserves the structure of the objects. Another name for a homomorphism of modules over R is an R-linear map.

A bijective module homomorphism is an isomorphism of modules, and the two modules are called *isomorphic*. Two isomorphic modules are identical for all practical purposes, differing solely in the notation for their elements.

The kernel of a module homomorphism $f: M \rightarrow N$ is the submodule of M consisting of all elements that are sent to zero by f. The isomorphism theorems familiar from groups and vector spaces are also valid for R-modules.

The left R-modules, together with their module homomorphisms, form a category, written as R-Mod. This is an abelian category.

Types of Modules

Finitely generated. An R-module M is finitely generated if there exist finitely many elements x_1, ..., x_n in M such that every element of M is a linear combination of those elements with coefficients from the ring R.

Cyclic. A module is called a cyclic module if it is generated by one element.

Free. A free R-module is a module that has a basis, or equivalently, one that is isomorphic to a direct sum of copies of the ring R. These are the modules that behave very much like vector spaces.

Projective. Projective modules are direct summands of free modules and share many of their desirable properties.

Injective. Injective modules are defined dually to projective modules.

Flat. A module is called flat if taking the tensor product of it with any exact sequence of R-modules preserves exactness.

Torsionless module. A module is called torsionless if it embeds into its algebraic dual.

Simple. A simple module S is a module that is not $\{0\}$ and whose only submodules are $\{0\}$ and S. Simple modules are sometimes called *irreducible*.

Semisimple. A semisimple module is a direct sum (finite or not) of simple modules. Historically these modules are also called *completely reducible*.

Indecomposable. An indecomposable module is a non-zero module that cannot be written as a direct sum of two non-zero submodules. Every simple module is indecomposable, but there are indecomposable modules which are not simple (e.g. uniform modules).

Faithful. A faithful module M is one where the action of each $r \neq 0$ in R on M is nontrivial (i.e. $r \cdot x \neq 0$ for some x in M). Equivalently, the annihilator of M is the zero ideal.

Torsion-free. A torsion-free module is a module over a ring such that 0 is the only element annihilated by a regular element (non zero-divisor) of the ring.

Noetherian. A Noetherian module is a module which satisfies the ascending chain condition on submodules, that is, every increasing chain of submodules becomes stationary after finitely many steps. Equivalently, every submodule is finitely generated.

Artinian. An Artinian module is a module which satisfies the descending chain condition on submodules, that is, every decreasing chain of submodules becomes stationary after finitely many steps.

Graded. A graded module is a module with a decomposition as a direct sum $M = \oplus_x M_x$ over a graded ring $R = \oplus_x R_x$ such that $R_x M_y \subset M_{x+y}$ for all x and y.

Uniform. A uniform module is a module in which all pairs of nonzero submodules have nonzero intersection.

Further Notions

Relation to Representation Theory

If M is a left R-module, then the *action* of an element r in R is defined to be the map $M \rightarrow M$ that sends each x to rx (or xr in the case of a right module), and is necessarily a group endomorphism of the abelian group $(M, +)$. The set of all group endomorphisms of M is denoted $\mathrm{End}_Z(M)$ and forms a ring under addition and composition, and sending a ring element r of R to its action actually defines a ring homomorphism from R to $\mathrm{End}_Z(M)$.

Such a ring homomorphism $R \rightarrow \mathrm{End}_Z(M)$ is called a *representation* of R over the abelian group M; an alternative and equivalent way of defining left R-modules is to say that a left R-module is an abelian group M together with a representation of R over it.

A representation is called *faithful* if and only if the map $R \rightarrow \mathrm{End}_Z(M)$ is injective. In terms of modules, this means that if r is an element of R such that $rx = 0$ for all x in M, then $r = 0$. Every abelian group is a faithful module over the integers or over some modular arithmetic Z/nZ.

Generalizations

Any ring R can be viewed as a preadditive category with a single object. With this understanding, a left R-module is nothing but a (covariant) additive functor from R to the category Ab of abelian groups. Right R-modules are contravariant additive functors. This suggests that, if C is any preadditive category, a covariant additive functor from C to Ab should be considered a generalized left module over C; these functors form a functor category C-Mod which is the natural generalization of the module category R-Mod.

Modules over *commutative* rings can be generalized in a different direction: take a ringed space (X, O_X) and consider the sheaves of O_X-module. These form a category O_X-Mod, and play an important role in modern algebraic geometry. If X has only a single point, then this is a module category in the old sense over the commutative ring $O_X(X)$.

One can also consider modules over a semiring. Modules over rings are abelian groups, but modules over semirings are only commutative monoids. Most applications of modules are still possible. In particular, for any semiring S the matrices over S form a semiring over which the tuples of elements from S are a module (in this generalized sense only). This allows a further generalization of the concept of vector space incorporating the semirings from theoretical computer science.

Over near-rings, one can consider near-ring modules, a nonabelian generalization of modules.

Algebraically Compact Module

In mathematics, especially in the area of abstract algebra known as module theory and in model theory, algebraically compact modules, also called pure-injective modules, are modules that have a certain "nice" property which allows the solution of infinite systems of equations in the module by finitary means. The solutions to these systems allow the extension of certain kinds of module homomorphisms. These algebraically compact modules are analogous to injective modules, where one can extend all module homomorphisms. All injective modules are algebraically compact, and the analogy between the two is made quite precise by a category embedding.

Definitions

Suppose R is a ring and M is a left R-module. Take two sets I and J, and for every i in I and j in J, an element r_{ij} of R such that, for every i in I, only finitely many r_{ij} are non-zero. Furthermore, take an element m_i of M for every i in I. These data describe a *system of linear equations* in M:

$$\sum_{j \in J} r_{ij} x_j = m_i \quad \text{for every } i \in I.$$

The goal is to decide whether this system has a *solution*, i.e. whether there exist elements x_j of M for every j in J such that all the equations of the system are simultaneously satisfied. (Note that we do not require that only finitely many of the x_j are non-zero here.)

Now consider such a system of linear equations, and assume that any subsystem consisting of only *finitely many* equations is solvable. (The solutions to the various subsystems may be different.) If every such "finitely-solvable" system is itself solvable, then we call the module M algebraically compact.

A module homomorphism $M \to K$ is called *pure injective* if the induced homomorphism between the tensor products $C \otimes M \to C \otimes K$ is injective for every right R-module C. The module M is pure-injective if any pure injective homomorphism $j : M \to K$ splits (i.e. there exists $f : K \to M$ with $fj = 1_M$).

It turns out that a module is algebraically compact if and only if it is pure-injective.

Examples

All modules with finitely many elements are algebraically compact.

Every vector space is algebraically compact (since it is pure-injective). More generally, every injective module is algebraically compact, for the same reason.

If R is an associative algebra with 1 over some field k, then every R-module with finite k-dimension is algebraically compact. This, together with that fact that all finite modules are algebraically compact, gives rise to the intuition that algebraically compact modules are those (possibly "large") modules which share the nice properties of "small" modules.

The Prüfer groups are algebraically compact abelian groups (i.e. Z-modules). The ring of p-adic integers for each prime p is algebraically compact as both a module over itself and a module over Z. The rational numbers are algebraically compact as a Z-module. Together with the indecomposable finite modules over Z, this is a complete list of indecomposable algebraically compact modules.

Many algebraically compact modules can be produced using the injective cogenerator Q/Z of abelian groups. If H is a *right* module over the ring R, one forms the (algebraic) character module H^* consisting of all group homomorphisms from H to Q/Z. This is then a left R-module, and the *-operation yields a faithful contravariant functor from right R-modules to left R-modules. Every module of the form H^* is algebraically compact. Furthermore, there are pure injective homomorphisms $H \to H^{**}$, natural in H. One can often simplify a problem by first applying the *-functor, since algebraically compact modules are easier to deal with.

Facts

The following condition is equivalent to M being algebraically compact:

- For every index set I, the addition map $M^{(I)} \to M$ can be extended to a module homomorphism $M^I \to M$ (here $M^{(I)}$ denotes the direct sum of copies of M, one for each element of I; M^I denotes the product of copies of M, one for each element of I).

Every indecomposable algebraically compact module has a local endomorphism ring.

Algebraically compact modules share many other properties with injective objects because of the following: there exists an embedding of R-Mod into a Grothendieck category G under which the algebraically compact R-modules precisely correspond to the injective objects in G.

Every R-module is elementary equivalent to an algebraically compact R-module and to a direct sum of indecomposable algebraically compact R-modules.

Field

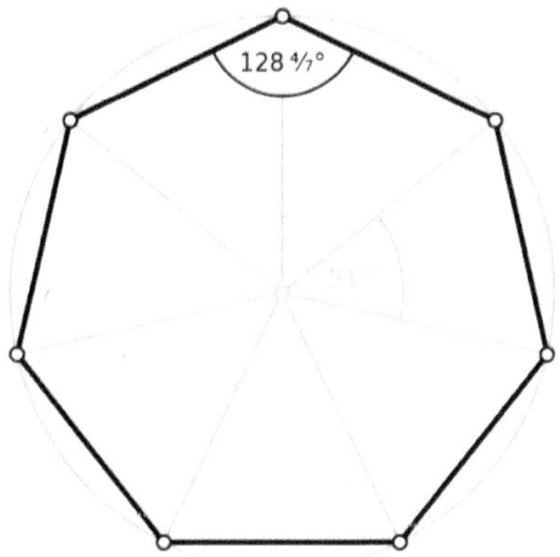

The regular heptagon cannot be constructed using compass and straightedge.
This can be proven using the field of constructible numbers.

In mathematics, a field is a set on which addition, subtraction, multiplication, and division are defined, and behave as when they are applied to rational and real numbers. A field is thus a fundamental algebraic structure, which is widely used in algebra, number theory and many other areas of mathematics.

The best known fields are the field of rational numbers and the field of real numbers. The field of complex numbers is also widely used, not only in mathematics, but also in many areas of science and engineering. Many other fields, such as fields of rational functions, algebraic function fields, algebraic number fields, and p-adic fields are commonly used and studied in mathematics, particularly in number theory and algebraic geometry. Most cryptographic protocols rely on finite fields, i.e., fields with finitely many elements.

The relation of two fields is expressed by the notion of a field extension. Galois theory, initiated by Évariste Galois in the 1830s, is devoted to understanding the symmetries of field extensions. Among other results, this theory shows that angle trisection and squaring the circle can not be done with a compass and straightedge. Moreover, it shows that quintic equations are algebraically unsolvable.

Fields serve as foundational notions in several mathematical domains. This includes different branches of analysis, which are based on fields with additional structure. Basic theorems in analysis hinge on the structural properties of the field of real numbers. Most importantly for algebraic purposes, any field may be used as the scalars for a vector space, which is the standard general context for linear algebra. Number fields, the siblings of the field of rational numbers, are studied in depth in number theory. Function fields can help describe properties of geometric objects.

Definition

Informally, a field is a set, along with two operations defined on that set: an addition operation written as $a + b$, and a multiplication operation written as $a \cdot b$, both of which behave similarly as they behave for rational numbers and real numbers, including the existence of an additive inverse $-a$ for all elements a, and of a multiplicative inverse b^{-1} for every nonzero element b. This allows us to consider also the so-called *inverse* operations of subtraction $a - b$, and division a / b, via defining:

$a - b = a + (-b)$,

$a / b = a \cdot b^{-1}$.

Classic Definition

Formally, a field is a set F together with two operations called *addition* and *multiplication*. An operation is a mapping that associates an element of the set to *every* pair of its elements. The result of the addition of a and b is called the *sum* of a and b and denoted $a + b$. Similarly, the result of the multiplication of a and b is called the *product* of a and b, and denoted ab or $a \cdot b$. These operations are required to satisfy the following properties, referred to as *field axioms*. In the following definitions, a, b and c are arbitrary elements of the field F.

- *Associativity* of addition and multiplication: $a + (b + c) = (a + b) + c$ and $a \cdot (b \cdot c) = (a \cdot b) \cdot c$.
- *Commutativity* of addition and multiplication: $a + b = b + a$ and $a \cdot b = b \cdot a$.
- *Additive* and *multiplicative identity*: there exist two different elements 0 and 1 in F such that $a + 0 = a$ and $a \cdot 1 = a$.
- *Additive inverses*: for every a in F, there exists an element in F, denoted $-a$, called *additive inverse* of a, such that $a + (-a) = 0$.
- *Multiplicative inverses*: for every $a \neq 0$ in F, there exists an element in F, denoted by a^{-1}, $1/a$, or $1/a$, called the *multiplicative inverse* of a, such that $a \cdot a^{-1} = 1$.
- *Distributivity* of multiplication over addition: $a \cdot (b + c) = (a \cdot b) + (a \cdot c)$.

This may be summarized by saying: a field has two operations, the addition and the multiplication; it is an abelian group under addition, with 0 as additive identity; the nonzero elements form an abelian group under multiplication, and the multiplication is distributive over addition.

Alternative Definition

Fields can also be defined in different, but equivalent ways. One can alternatively define a field by four binary operations (add, subtract, multiply, divide), and their required properties. Division by zero is, by definition, excluded. In order to avoid existential quantifiers, fields can be defined by two binary operations (addition and multiplication), two unary operations (yielding the additive and multiplicative inverses, respectively), and two nullary operations (the constants 0 and 1). These operations are then subject to the conditions above. This approach avoids existential quantifiers, which is important in constructive mathematics and computing.

Examples

Rational Numbers

Rational numbers have been widely used a long time before the elaboration of the concept of field. They are numbers that can be written as fractions a/b, where a and b are integers, and $b \neq 0$. The additive inverse of such a fraction is $-a/b$, and the multiplicative inverse (provided that $a \neq 0$) is b/a, which can be seen as follows:

$$\frac{b}{a}\cdot\frac{a}{b}=\frac{ba}{ab}=1.$$

The abstractly required field axioms reduce to standard properties of rational numbers. For example, the law of distributivity can be proven as follows:

$$\begin{aligned}
&\frac{a}{b}\cdot\left(\frac{c}{d}+\frac{e}{f}\right)\\
&=\frac{a}{b}\cdot\left(\frac{c}{d}\cdot\frac{f}{f}+\frac{e}{f}\cdot\frac{d}{d}\right)\\
&=\frac{a}{b}\cdot\left(\frac{cf}{df}+\frac{ed}{fd}\right)=\frac{a}{b}\cdot\frac{cf+ed}{df}\\
&=\frac{a(cf+ed)}{bdf}=\frac{acf}{bdf}+\frac{aed}{bdf}=\frac{ac}{bd}+\frac{ae}{bf}\\
&=\frac{a}{b}\cdot\frac{c}{d}+\frac{a}{b}\cdot\frac{e}{f}.
\end{aligned}$$

Real and Complex Numbers

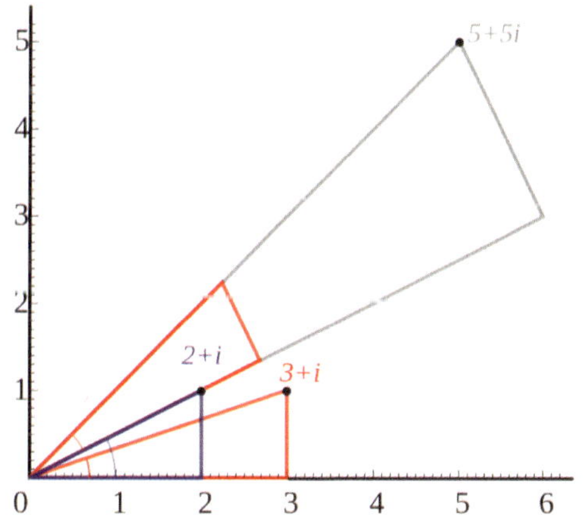

The multiplication of complex numbers can be visualized geometrically by rotations and scalings.

The real numbers R, with the usual operations of addition and multiplication, also form a field. The complex numbers C consist of expressions

$a + bi$, with a, b real,

where i is the imaginary unit, i.e., a (non-real) number satisfying $i^2 = -1$. Addition and multiplication of real numbers are defined in such a way that expressions of this type satisfy all field axioms and thus hold for C. For example, the distributive law enforces

$$(a + bi)\cdot(c + di) = ac + bci + adi + bdi^2 = ac-bd + (bc + ad)i.$$

It is immediate that this again an expression of the above type, and so the complex numbers form a field. Complex numbers can be geometrically represented as points in the plane, with Cartesian coordinates given by the real numbers of their describing expression, or as the arrows from the origin to these points, specified by their length and an angle enclosed with some distinct direction. Addition then corresponds to combining the arrows to the intuitive parallelogram (adding the Cartesian coordinates), and the multiplication is –less intuitively– combining rotating and scaling of the arrows (adding the angles and multiplying the lengths). The fields of real and complex numbers are used throughout mathematics, physics, engineering, statistics, and many other scientific disciplines.

Constructible Numbers

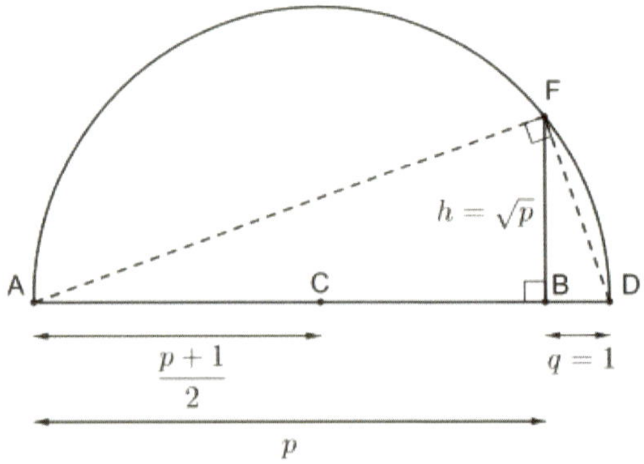

The geometric mean theorem asserts that $h^2 = pq$. Choosing $q = 1$ allows construction of the square root of a given constructible number p. Construct the segments AB, BD, and a semicircle over AD (center at the midpoint C), which intersects the perpendicular line through B in a point F, exactly $h = \sqrt{p}$ apart from B.

In antiquity, several geometric problems concerned the (in)feasibility of constructing certain numbers with compass and straightedge. For example, it was unknown to the Greeks that it is in general impossible to trisect a given angle. These problems can be settled using the field of constructible numbers. Real constructible numbers are, by definition, lengths of line segments that can be constructed from the points 0 and 1 in finitely many steps using only compass and straightedge. These numbers, endowed with the field operations of real numbers, restricted to the constructible numbers, form a field, which properly includes the field Q of rational numbers. The illustration shows the construction of square roots of constructible numbers, not necessarily contained within Q.

Not all real numbers are constructible. It can be shown that $\sqrt[3]{2}$ is not a constructible number, which implies that it is impossible to construct with compass and straightedge the length of the side of a cube with volume 2, another problem posed by the ancient Greeks.

A field with four elements

Addition

+	*O*	*I*	*A*	*B*
O	*O*	*I*	A	B
I	*I*	*O*	B	A
A	A	B	O	I
B	B	A	I	O

Multiplication

·	*O*	*I*	*A*	*B*
O	*O*	*O*	O	O
I	*O*	*I*	A	B
A	O	A	B	I
B	O	B	I	A

In addition to familiar number systems such as the rationals, there are other, less immediate examples of fields. The following example is a field consisting of four elements called O, I, A, and B. The notation is chosen such that O plays the role of the additive identity element (denoted 0 in the axioms above), and I is the multiplicative identity (denoted 1 in the axioms above). The field axioms can be verified by using some more field theory, or by direct computation. For example,

$$A \cdot (B + A) = A \cdot I = A, \text{ which equals } A \cdot B + A \cdot A = I + B = A, \text{ as required by the distributivity.}$$

This field is called a finite field with four elements, and is denoted F_4 or GF(4). The subset consisting of O and I (highlighted in red in the tables at the right) is also a field, known as the *binary field* F_2 or GF(2). In the context of computer science and Boolean algebra, O and I are often denoted respectively by *false* and *true*, the addition is then denoted XOR (exclusive or), and the multiplication is denoted AND. In other words, the structure of the binary field is the basic structure that allows computing with bits.

Elementary Notions

In this section, F denotes an arbitrary field and a and b are arbitrary elements of F.

Consequences of the Definition

One has $a \cdot 0 = 0$ and $-a = (-1) \cdot a$. In particular, one may deduce the additive inverse of every element as soon as one knows -1.

If $ab = 0$ then a or b must be 0. Indeed, if $a \neq 0$, then $0 = a^{-1} \cdot 0 = a^{-1}(ab) = (a^{-1}a)b = b$. This means that every field is an integral domain.

The Additive and the Multiplicative Group of a Field

The axioms of a field F imply that it is an abelian group under addition. This group is called the additive group of the field, and is sometimes denoted by $(F, +)$ when denoting it simply as F could be confusing.

Similarly, the *nonzero* elements of F form an abelian group under multiplication, called the multiplicative group, and denoted by $(F \setminus \{0\}, \cdot)$ or just $F \setminus \{0\}$ or F^*.

A field may thus be defined as set F equipped with two operations denoted as an addition and a multiplication such that F is an abelian group under addition, $F \setminus \{0\}$ is an abelian group under multiplication (where 0 is the identity element of the addition), and multiplication is distributive over addition. Some elementary statements about fields can therefore be obtained by applying general facts of groups. For example, the additive and multiplicative inverses $-a$ and a^{-1} are uniquely determined by a.

The requirement $1 \neq 0$ follows, because 1 is the identity element of a group that does not contain 0. Thus, the trivial ring, consisting of a single element, is not a field.

Every finite subgroup of the multiplicative group of a field is cyclic.

Characteristic

In addition to the multiplication of two elements of F, it is possible to define the product $n \cdot a$ of an arbitrary element a of F by a positive integer n to be the n-fold sum

$$a + a + \ldots + a \text{ (which is an element of } F.)$$

If there is no positive integer such that

$$n \cdot 1 = 0,$$

then F is said to have characteristic 0. For example, Q has characteristic 0 since no positive integer n is zero. Otherwise, if there *is* a positive integer n satisfying this equation, the smallest such positive integer can be shown to be a prime number. It is usually denoted by p and the field is said to have characteristic p then. For example, the field F_4 has characteristic 2 since (in the notation of the above addition table) I + I = O.

If F has characteristic p, then $p \cdot a = 0$ for all a in F. This implies that

$$(a + b)^p = a^p + b^p,$$

since all other binomial coefficients appearing in the binomial formula are divisible by p. Here, $a^p := a \cdot a \cdot \ldots \cdot a$ (p factors) is the p-th power, i.e., the p-fold product of the element a. Therefore, the Frobenius map

$$\text{Fr}: F \to F, x \longmapsto x^p$$

is compatible with the addition in F (and also with the multiplication), and is therefore a field homomorphism. The existence of this homomorphism makes fields in characteristic p quite different from fields of characteristic 0.

Subfields and Prime Fields

A *subfield* E of a field F is a subset of F that is a field with respect to the field operations of F. Equivalently E is a subset of F that contains 1, and is closed under addition, multiplication, additive inverse and multiplicative inverse of a nonzero element. This means that $1 \in E$, that for all $a, b \in E$ both $a + b$ and $a \cdot b$ are in E, and that for all $a \neq 0$ in E, both $-a$ and $1/a$ are in E.

Field homomorphisms are maps $f: E \to F$ between two fields such that $f(e_1 + e_2) = f(e_1) + f(e_2)$, $f(e_1e_2) = f(e_1)f(e_2)$, and $f(1_E) = 1_F$, where e_1 and e_2 are arbitrary elements of E. All field homomorphisms are injective. If f is also surjective, it is called an isomorphism (or the fields E and F are called isomorphic).

A field is called a prime field if it has no proper (i.e., strictly smaller) subfields. Any field F contains a prime field. If the characteristic of F is p (a prime number), the prime field is isomorphic to the finite field F_p introduced below. Otherwise the prime field is isomorphic to Q.

Finite Fields

Finite fields (also called *Galois fields*) are fields with finitely many elements, whose number is also referred to as the order of the field. The above introductory example F_4 is a field with four elements. Its subfield F_2 is the smallest field, because by definition a field has at least two distinct elements $1 \neq 0$.

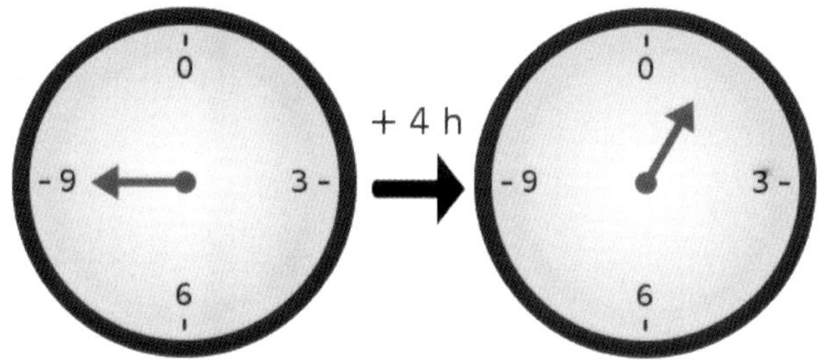

In modular arithmetic modulo 12, $9 + 4 = 1$ since $9 + 4 = 13$ in Z, which divided by 12 leaves remainder 1. Since 12 is not a prime, Z/12Z is not a field, though.

The simplest finite fields, with prime order, are most directly accessible using modular arithmetic. For a fixed positive integer n, arithmetic "modulo n" means to work with the numbers

$$Z/nZ = \{0, 1, ..., n - 1\}.$$

The addition and multiplication on this set are done by performing the operation in question in the set Z of integers, dividing by n and taking the remainder as result. This construction yields a field precisely if n is a prime number. For example, taking the prime $n = 2$ results in the above-mentioned field F_2. For $n = 4$ and more generally, for any composite number (i.e., any number n which can be expressed as a product $n = r{\cdot}s$ of two strictly smaller natural numbers), Z/nZ is not a field: the product of two non-zero elements is zero since $r{\cdot}s = 0$ in Z/nZ, which, as was explained above, prevents Z/nZ from being a field. The field Z/pZ with p elements (p being prime) constructed in this way is usually denoted by F_p.

Every finite field F has $q = p^n$ elements, where p is prime and $n \geq 1$. This statement holds since F may be viewed as a vector space over its prime field. The dimension of this vector space is necessarily finite, say n, which implies the asserted statement.

A field with $q = p^n$ elements can be constructed as the splitting field of the polynomial

$$f(x) = x^q - x.$$

Such a splitting field is an extension of F_p in which the polynomial f has q zeros. This means f has as many zeros as possible since the degree of f is q. For $q = 2^2 = 4$, it can be checked case by case using the above multiplication table that all four elements of F_4 satisfy the equation $x^4 = x$, so they are zeros of f. By contrast, in F_2, f has only two zeros (namely 0 and 1), so f does not split into linear factors in this smaller field. Elaborating further on basic field-theoretic notions, it can be shown that two finite fields with the same order are isomorphic. It is thus customary to speak of *the* finite field with q elements, denoted by F_q or GF(q).

Constructing Fields

Constructing Fields from Rings

A commutative ring is a set, equipped with an addition and multiplication operation, satisfying all the axioms of a field, except for the existence of multiplicative inverses a^{-1}. For example, the integers Z form a commutative ring, but not a field: the reciprocal of an integer n is not itself an integer, unless $n = \pm 1$.

In the hierarchy of algebraic structures fields can be characterized as the commutative rings R in which every nonzero element is a unit (which means every element is invertible). Similarly, fields are the commutative rings with precisely two distinct ideals, (0) and R. Fields are also precisely the commutative rings in which (0) is the only prime ideal.

Given a commutative ring R, there are two ways to construct a field related to R, i.e., two ways of modifying R such that all nonzero elements become invertible: forming the field of fractions, and forming residue fields. The field of fractions of Z is Q, the rationals, while the residue fields of Z are the finite fields F_p.

Field of Fractions

Given an integral domain R, its field of fractions $Q(R)$ is built with the fractions of two elements of R exactly as Q is constructed from the integers. More precisely, the elements of $Q(R)$ are the fractions a/b where a and b are in R, and $b \neq 0$. Two fractions a/b and c/d are equal if and only if $ad = bc$. The operation on the fractions work exactly as for rational numbers. For example,

$$\frac{a}{b}+\frac{c}{d}=\frac{ad+bc}{bd}.$$

It is straightforward to show that, if the ring is an integral domain, the set of the fractions form a field.

The field $F(x)$ of the rational fractions over a field (or an integral domain) F is the field of fractions of the polynomial ring $F[x]$. The field $F((x))$ of Laurent series

$$\sum_{i=k}^{\infty} a_i x^i \ (k \in \mathbb{Z}, a_i \in F)$$

over a field F is the field of fractions of the ring $F[[x]]$ of formal power series (in which $k \geq 0$). Since any Laurent series is a fraction of a power series divided by a power of x (as opposed to an arbitrary power series), the representation of fractions is less important in this situation, though.

Residue Fields

In addition the field of fractions, which embedds R injectively into a field, a field can be obtained from a commutative ring R by means of a surjective map onto a field F. Any field obtained in this way is a quotient R / m, where m is a maximal ideal of R. If R has only one maximal ideal m, this field is called the residue field of R.

The ideal generated by a single polynomial f in the polynomial ring $R = E[X]$ (over a field E) is maximal if and only if f is irreducible in E, i.e., if f can not be expressed as the product of two polynomials in $E[X]$ of smaller degree. This yields a field

$$F = E[X] / (p(X)).$$

This field F contains an element x (namely the residue class of X) which satisfies the equation

$$f(x) = 0.$$

For example, C is obtained from R by adjoining the imaginary unit symbol i, which satisfies $f(i) = 0$, where $f(X) = X^2 + 1$. Moreover, f is irreducible over R, which implies that the map that sends a polynomial $f(X) \in R[X]$ to $f(i)$ yields an isomorphism

$$\mathbf{R}[X]/(X^2+1) \xrightarrow{\cong} \mathbf{C}.$$

Constructing Fields within a Bigger Field

Fields can be constructed inside a given bigger container field. Suppose given a field E, and a field F containing E as a subfield. For any element x of F, there is a smallest subfield of F containing E and x, called the subfield of F generated by x and denoted $E(x)$. The passage from E to $E(x)$ is referred to by *adjoining an element* to E. More generally, for a subset $S \subset F$, there is a minimal subfield of F containing E and S, denoted by $E(S)$.

The compositum of two subfields E and E' of some field F is the smallest subfield of F containing both E and E'. The compositum can be used to construct the biggest subfield of F satisfying a certain property, for example the biggest subfield of F, which is, in the language introduced below, algebraic over E.

Field Extensions

The notion of a subfield $E \subset F$ can also be regarded from the opposite point of view, by referring to F being a *field extension* (or just extension) of E, denoted by

$$F / E,$$

and read "F over E".

A basic datum of a field extension is its degree $[F : E]$, i.e., the dimension of F as an E-vector space. It satisfies the formula

$$[G : E] = [G : F]\,[F : E].$$

Extensions whose degree is finite are referred to as finite extensions. The extensions C / R and F_4 / F_2 are of degree 2, whereas R / Q is an infinite extension.

Algebraic Extensions

A pivotal notion in the study of field extensions F / E are algebraic elements. An element $x \in F$ is *algebraic* over E if it is a root of a polynomial with coefficients in E, that is, if it satisfies a polynomial equation

$$e_n x^n + e_{n-1} x^{n-1} + \cdots + e_1 x + e_0 = 0,$$

with e_n, ..., e_0 in E, and $e_n \neq 0$. For example, the imaginary unit i in C is algebraic over R, and even over Q, since it satisfies the equation

$$i^2 + 1 = 0.$$

A field extension in which every element of F is algebraic over E is called an algebraic extension. Any finite extension is necessarily algebraic, as can be deduced from the above multiplicativity formula.

The subfield $E(x)$ generated by an element x, as above, is an algebraic extension of E if and only if x is an algebraic element. That is to say, if x is algebraic, all other elements of $E(x)$ are necessarily algebraic as well. Moreover, the degree of the extension $E(x) / E$, i.e., the dimension of $E(x)$ as an E-vector space, equals the minimal degree n such that there is a polynomial equation involving x, as above. If this degree is n, then the elements of $E(x)$ have the form

$$\sum_{k=0}^{n-1} a_k x^k, \quad a_k \in E.$$

For example, the field Q(i) of Gaussian rationals is the subfield of C consisting of all numbers of the form $a + bi$ where both a and b are rational numbers: summands of the form i^2 (and similarly for higher exponents) don't have to be considered here, since $a + bi + ci^2$ can be simplified to $a - c + bi$.

Transcendence Bases

The above-mentioned field of rational fractions $E(X)$, where X is an indeterminate, is not an algebraic extension of E since there is no polynomial equation with coefficients in E whose zero is X. Elements, such as X, which are not algebraic are called transcendental. Informally speaking, the indeterminate X and its powers do not interact with elements of E. A similar construction can be carried out with a set of indeterminates, instead of just one.

Once again, the field extension $E(x) / E$ discussed above is a key example: if x is not algebraic (i.e., x is not a root of a polynomial with coefficients in E), then $E(x)$ is isomorphic to $E(X)$. This isomorphism is obtained by substituting x to X in rational fractions.

A subset S of a field F is a transcendence basis if it is algebraically independent (don't satisfy any polynomial relations) over E and if F is an algebraic extension of $E(S)$. Any field extension F / E has a transcendence basis. Thus, field extensions can be split into ones of the form $E(S) / E$ (purely transcendental extensions) and algebraic extensions.

Closure Operations

A field is algebraically closed if it does not have any strictly bigger algebraic extensions or, equivalently, if any polynomial equation

$$f_n x^n + f_{n-1} x^{n-1} + \cdots + f_1 x + f_0 = 0, \text{ with coefficients } f_n, \ldots, f_0 \in F, n > 0,$$

has a solution $x \in F$. By the fundamental theorem of algebra, C is algebraically closed, i.e., *any* polynomial equation with complex coefficients has a complex solution. The rational and the real numbers are *not* algebraically closed since the equation

$$x^2 + 1 = 0$$

does not have any rational or real solution. A field containing F is called an *algebraic closure* of F if it is algebraic over F (roughly speaking, not too big compared to F) and is algebraically closed (big enough to contain solutions of all polynomial equations).

By the above, C is an algebraic closure of R. The situation that the algebraic closure is a finite extension of the field F is quite special: by the Artin-Schreier theorem, the degree of this extension is necessarily 2, and F is elementarily equivalent to R. Such fields are also known as real closed fields.

Any field F has an algebraic closure, which is moreover unique up to (non-unique) isomorphism. It is commonly referred to as *the* algebraic closure and denoted $\overline{\mathrm{F}}$. For example, the algebraic closure $\overline{\mathrm{Q}}$ of Q is called the field of algebraic numbers. The field $\overline{\mathrm{F}}$ is usually rather implicit since its construction requires the ultrafilter lemma, a set-theoretic axiom that is weaker than the axiom of choice. In this regard, the algebraic closure of F_q, is exceptionally simple. It is the union of the finite fields containing F_q (the ones of order q^n). For any algebraically closed field F of characteristic 0, the algebraic closure of the field $F((t))$ of Laurent series is the field of Puiseux series, obtained by adjoining roots of t.

Fields with Additional Structure

Since fields are ubiquitous in mathematics and beyond, several refinements of the concept have been adapted to the needs of particular mathematical areas.

Ordered Fields

A field F is called an *ordered field* if any two elements can be compared, so that $x + y \geq 0$ and $xy \geq 0$ whenever $x \geq 0$ and $y \geq 0$. For example, the reals form an ordered field, with the usual ordering $\geq$. The Artin-Schreier theorem states that a field can be ordered if and only if it is a formally real field, which means that any quadratic equation

$$x_1^2 + x_2^2 + \ldots + x_n^2 = 0$$

only has the solution $x_1 = x_2 = \ldots = x_n = 0$. The set of all possible orders on a fixed field F is isomorphic to the set of ring homomorphisms from the Witt ring W(F) of quadratic forms over F, to Z.

An Archimedean field is an ordered field such that for each element there exists a finite expression

$$1 + 1 + \cdots + 1$$

whose value is greater than that element, that is, there are no infinite elements. Equivalently, the field contains no infinitesimals (elements smaller than all rational numbers); or, yet equivalent, the field is isomorphic to a subfield of R.

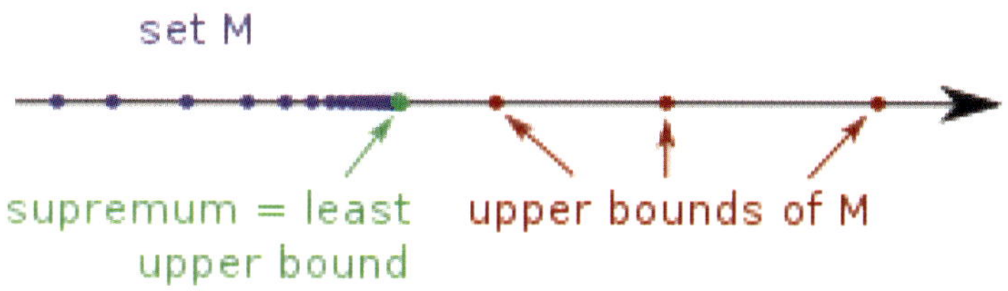

Each bounded real set has a least upper bound.

An ordered field is Dedekind-complete if all upper bounds, lower bounds and limits, which should exist, do exist. More formally, each bounded subset of F is required to have a least upper bound. Any complete field is necessarily Archimedean, since in any non-Archimedean field there is neither a greatest infinitesimal nor a least positive rational, whence the sequence 1/2, 1/3, 1/4, ..., every element of which is greater than every infinitesimal, has no limit.

Since every proper subfield of the reals also contains such gaps, R is the unique complete ordered field, up to isomorphism. Several foundational results in calculus follow directly from this characterization of the reals.

The hyperreals R^* form an ordered field that is not Archimedean. It is an extension of the reals obtained by including infinite and infinitesimal numbers. These are larger, respectively smaller than any real number. The hyperreals form the foundational basis of non-standard analysis.

Topological Fields

Another refinement of the notion of a field is a topological field, in which the set F is a topological space, such that all operations of the field (addition, multiplication, the maps $a \mapsto -a$ and $a \mapsto a^{-1}$) are continuous maps with respect to the topology of the space. The topology of all the fields discussed below is induced from a metric, i.e., a function

$$d : F \times F \to R,$$

that measures a *distance* between any two elements of F.

The completion of F is another field in which, informally speaking, the "gaps" in the original field F are filled, if there are any. For example, any irrational number x, such as $x = \sqrt{2}$, is a "gap" in the sense that it is a real number that can be approximated arbitrarily closely by rational numbers p/q, in the sense that distance of x and p/q given by the absolute value $|x - p/q|$ is as small as desired. The following table lists some examples of this construction. The fourth column shows an example of a zero sequence, i.e., a sequence whose limit (for $n \to \infty$) is zero.

Field	Metric	Completion	zero sequence
Q	$\|x - y\|$ (usual absolute value)	**R**	$1/n$
Q	obtained using the p-adic valuation, for a prime number p	$\mathbf{Q}_p$ p-adic numbers	p^n
$F(t)$ (F any field)	obtained using the t-adic valuation	$F((t))$	t^n

The field Q_p is used in number theory and p-adic analysis. The algebraic closure Q_p carries a unique norm extending the one on Q_p, but is not complete. The completion of this algebraic closure, however, is algebraically closed. Because of its rough analogy to the complex numbers, it is called the field of complex p-adic numbers and is denoted by C_p.

Local Fields

The following topological fields are called *local fields*:

- finite extensions of Q_p (local fields of characteristic zero).

- finite extensions of $F_p((t))$, the field of Laurent series over F_p (local fields of characteristic p).

These two types of local fields share some fundamental similarities. In this relation, the elements $p \in Q_p$ and $t \in F_p((t))$ (referred to as uniformizer) correspond to each other. The first manifestation of this is at an elementary level: the elements of both fields can be expressed as power series in the uniformizer, with coefficients in F_p. (However, since the addition in Q_p is done using carrying, which is not the case in $F_p((t))$, these fields are not isomorphic.) The following facts show that this superficial similarity goes much deeper:

- Any first order statement that is true for almost all Q_p is also true for almost all $F_p((t))$. An application of this is the Ax-Kochen theorem describing zeros of homogeneous polynomials in Q_p.
- Tamely ramified extensions of both fields are in bijection to one another.
- Adjoining arbitrary p-power roots of p (in Q_p), respectively of t (in $F_p((t))$), yields (infinite) extensions of these fields known as perfectoid fields. Strikingly, the Galois groups of these two fields are isomorphic, which is the first glimpse of a remarkable parallel between these two fields:

$$\mathrm{Gal}(\mathbf{Q}_p(p^{1/p^\infty})) \cong \mathrm{Gal}(\mathbf{F}_p((t))(t^{1/p^\infty})).$$

Differential Fields

Differential fields are fields equipped with a derivation, i.e., allow to take derivatives of elements in the field. For example, the field $R(X)$, together with the standard derivative of polynomials forms a differential field. These fields are central to differential Galois theory, a variant of Galois theory dealing with linear differential equations.

Galois Theory

Galois theory studies algebraic extensions of a field by studying the symmetry in the arithmetic operations of addition and multiplication. An important notion in this area are finite Galois extensions F / E, which are, by definition, those that are separable and normal. The primitive element theorem shows that finite separable extensions are necessarily simple, i.e., of the form

$$F = E[X] / f(X),$$

where f is an irreducible polynomial (as above). For such an extension, being normal and separable means that all zeros of f are contained in F and that f has only simple zeros. The latter condition is always satisfied if E has characteristic 0.

For a finite Galois extension, the Galois group $\mathrm{Gal}(F/E)$ is the group of field automorphisms of F that are trivial on E (i.e., the bijections $\sigma : F \to F$ that preserve addition and multiplication and that send elements of E to themselves). The importance of this group stems from the fundamental theorem of Galois theory, which constructs an explicit one-to-one correspondence between the set of subgroups of $\mathrm{Gal}(F/E)$ and the set of intermediate extensions of the extension F/E. By means of this correspondence, group-theoretic properties translate into facts about fields. For example, if the Galois group of a Galois extension as above is not solvable (can not be built from abelian groups), then the zeros of f can *not* be expressed in terms of addition, multiplication, and radicals, i.e., expressions involving $\sqrt[n]{\ }$. For example, the symmetric groups S_n is not solvable for $n \geq 5$. Consequently, as can be shown, the zeros of the following polynomials are not expressible by sums, products, and radicals. For the latter polynomial, this fact is known as the Abel–Ruffini theorem:

$f(X) = X^5 - 4X + 2$ (and $E = Q$),

$f(X) = X^n + a_{n-1}X^{n-1} + \ldots + a_0$ (where f is regarded as a polynomial in $E(a_0, \ldots, a_{n-1})$, for some indeterminates a_i, E is any field, and $n \geq 5$).

The tensor product of fields is not usually a field. For example, a finite extension F / E of degree n is a Galois extension if and only if there is an isomorphism of F-algebras

$F \otimes_E F \cong F^n$.

This fact is the beginning of Grothendieck's Galois theory, a far-reaching extension of Galois theory applicable to algebro-geometric objects.

Invariants of Fields

Basic invariants of a field F include the characteristic and the transcendence degree of F over its prime field. The latter is defined as the maximal number of elements in F that are algebraically independent over the prime field. Two algebraically closed fields E and F are isomorphic precisely if these two data agree. This implies that any two uncountable algebraically closed fields of the same cardinality and the same characteristic are isomorphic. For example, $\overline{Q}_p$, C_p and C are isomorphic (but *not* isomorphic as topological fields).

Model Theory of Fields

In model theory, a branch of mathematical logic, two fields E and F are called elementarily equivalent if every mathematical statement that is true for E is also true for F and conversely. The mathematical statements in question are required to be first-order sentences (involving 0, 1, the addition and multiplication). A typical example is

$\varphi(E)$ = "for any $n > 0$, any polynomial of degree n in E has a zero in E" (which amounts to saying that E is algebraically closed).

The Lefschetz principle states that C is elementarily equivalent to any algebraically closed field F of characteristic zero. Moreover, any fixed statement φ holds in C if and only if it holds in any algebraically closed field of sufficiently high characteristic.

If U is an ultrafilter on a set I, and F_i is a field for every i in I, the ultraproduct of the F_i with respect to U is a field. It is denoted by

$\text{ulim}_{i\to\infty} F_i$,

since it behaves in several ways as a limit of the fields F_i: Łoś›s theorem states that any first order statement that holds for all but finitely many F_i, also holds for the ultraproduct. Applied to the above sentence φ, this shows that there is an isomorphism

$$\mathrm{ulim}_{p\to\infty} \overline{\mathbf{F}}_p \cong \mathbf{C}.$$

The Ax–Kochen theorem mentioned above also follows from this and an isomorphism of the ultraproducts (in both cases over all primes p)

$$\mathrm{ulim}_p \mathrm{Q}_p \cong \mathrm{ulim}_p \mathrm{F}_p((t)).$$

In addition, model theory also studies the logical properties of various other types of fields, such as real closed fields or exponential fields (which are equipped with an exponential function exp : $F \to F^{\times}$).

The Absolute Galois Group

For fields that are not algebraically closed (or not separably closed), the absolute Galois group Gal(F) is fundamentally important: extending the case of finite Galois extensions outlined above, this group governs *all* finite separable extensions of F. By elementary means, the group Gal(F_q) can be shown to be the Prüfer group, the profinite completion of Z. This statement subsumes the fact that the only algebraic extensions of Gal(F_q) are the fields Gal($\mathrm{F}_q{}^n$) for $n > 0$, and that the Galois groups of these finite extensions are given by

$$\mathrm{Gal}(\mathrm{F}_q{}^n / \mathrm{F}_q) = \mathrm{Z}/n\mathrm{Z}.$$

A description in terms of generators and relations is also known for the Galois groups of p-adic number fields (finite extensions of Q_p).

Representations of Galois groups and of related groups such as the Weil group are fundamental in many branches of arithmetic, such as the Langlands program. The cohomological study of such representations is done using Galois cohomology. For example, the Brauer group, which is classically defined as the group of central simple F-algebras, can be reinterpreted as a Galois cohomology group, namely

$$\mathrm{Br}(F) = \mathrm{H}^2(F, \mathrm{G}_\mathrm{m}).$$

K-theory

Milnor K-theory is defined as

$$K_n^M(F) = F^{\times} \otimes \cdots \otimes F^{\times} / \langle x \otimes (1-x) \mid x \in F \setminus \{0,1\} \rangle.$$

The norm residue isomorphism theorem, proved around 2000 by Vladimir Voevodsky, relates this to Galois cohomology by means of an isomorphism

$$K_n^M(F)/p = H^n(F, \mu_l^{\otimes n}).$$

Algebraic K-theory is related to the group of invertible matrices with coefficients the given field.

For example, the process of taking the determinant of an invertible matrix leads to an isomorphism $K_1(F) = F^\times$. Matsumoto's theorem shows that $K_2(F)$ agrees with $K_2^M(F)$. In higher degrees, K-theory diverges from Milnor K-theory and remains hard to compute in general.

Applications

Linear Algebra and Commutative Algebra

If $a \neq 0$, then the equation

$$ax = b$$

has a unique solution x in F, namely $x = b/a$. This observation, which is an immediate consequence of the definition of a field, is the essential ingredient used to show that any vector space has a basis. Roughly speaking, this allows choosing a coordinate system in any vector space, which is of central importance in linear algebra both from a theoretical point of view, and also for practical applications.

Modules (the analogue of vector spaces) over most rings, including the ring Z of integers, have a more complicated structure. A particular situation arises when a ring R is a vector space over a field F in its own right. Such rings are called F-algebras and are studied in depth in the area of commutative algebra. For example, Noether normalization asserts that any finitely generated F-algebra is closely related to (more precisely, finitely generated as a module over) a polynomial ring $F[x_1, ..., x_n]$.

Finite Fields: Cryptography and Coding Theory

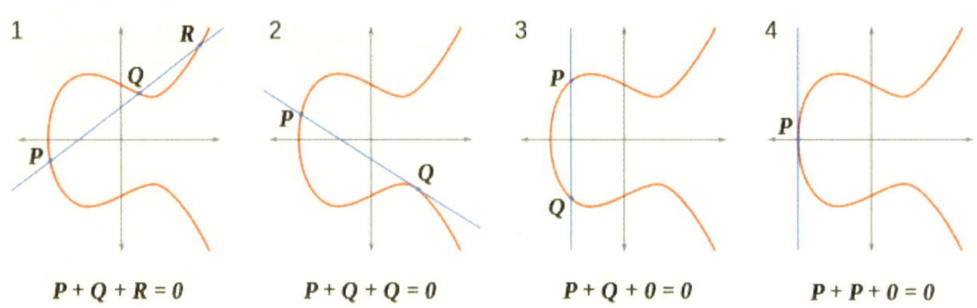

The sum of three points P, Q, and R on an elliptic curve E (red) is zero if there is a line (blue) passing through these points.

A widely applied cryptographic routine uses the fact that discrete exponentiation, i.e., computing

$$a^n = a \cdot a \cdot ... \cdot a \text{ (}n\text{ factors, for an integer } n \geq 1)$$

in a (large) finite field F_q can be performed much more efficiently than the discrete logarithm, which is the inverse operation, i.e., determining the solution n to an equation

$$a^n = b.$$

In elliptic curve cryptography, the multiplication in a finite field is replaced by the operation of adding points on an elliptic curve, i.e., the solutions of an equation of the form

$$y^2 = x^3 + ax + b.$$

Finite fields are also used in coding theory and combinatorics.

Geometry: Field of Functions

A compact Riemann surface of genus two (two handles).
The genus can be read off the field of meromorphic functions on the surface.

Functions on a suitable topological space X into a field k can be added and multiplied pointwise, e.g., the product of two functions is defined by the product of their values within the domain:

$$(f \cdot g)(x) = f(x) \cdot g(x).$$

This makes these functions a k-commutative algebra.

For having a *field* of functions, one must consider algebras of functions that are integral domains. In this case the ratios of two functions, i.e., expressions of the form

$$\frac{f(x)}{g(x)},$$

form a field, called field of functions.

This occurs in two main cases. When X is a complex manifold X. In this case, one considers the algebra of holomorphic functions, i.e., complex differentiable functions. Their ratios form the field of meromorphic functions on X.

The function field of an algebraic variety X (a geometric object defined as the common zeros of polynomial equations) consists of ratios of regular functions, i.e., ratios of polynomial functions on the variety. The function field of the n-dimensional space over a field k is $k(x_1, ..., x_n)$, i.e., the field consisting of ratios of polynomials in n indeterminates. The function field of X is the same as the one of any open dense subvariety. In other words, the function field is insensitive to replacing X by a (slightly) smaller subvariety.

The function field is invariant under isomorphism and birational equivalence of varieties. It is therefore an important tool for the study of abstract algebraic varieties and for the classification of algebraic varieties. For example, the dimension, which equals the transcendence degree of $k(X)$, is invariant under birational equivalence. For curves (i.e., the dimension is one), the function field $k(X)$ is very close to X: if X is smooth and proper (the analogue of being compact), X can be reconstructed, up to isomorphism, from its field of functions. In higher dimension the function field remembers less, but still decisive information about X. The study of function fields and their geo-metric meaning in higher dimensions is referred to as birational geometry. The minimal model program attempts to identify the simplest (in a certain precise sense) algebraic varieties with a prescribed function field.

Number Theory: Global Fields

Global fields are in the limelight in algebraic number theory and arithmetic geometry. They are, by definition, number fields (finite extensions of Q) or function fields over F_q (finite extensions of $F_q(t)$). As for local fields, these two types of fields share several similar features, even though they are of characteristic 0 and positive characteristic, respectively. This function field analogy can help to shape mathematical expectations, often first by understanding questions about function fields, and later treating the number field case. The latter is often more difficult. For example, the Riemann hypothesis concerning the zeros of the Riemann zeta function can be regarded as being parallel to the Weil conjectures.

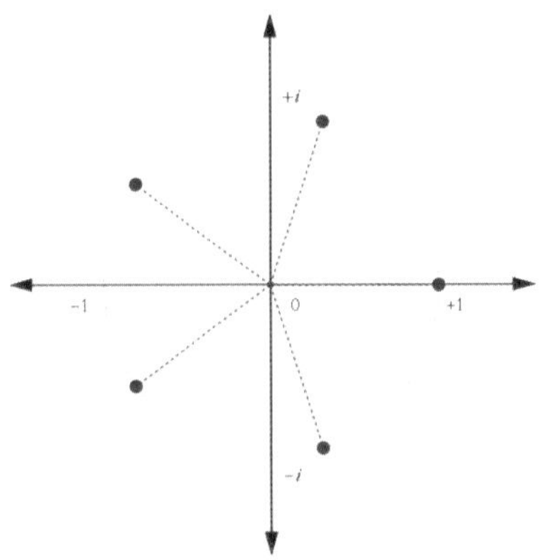

The fifth roots of unity form a regular pentagon.

Cyclotomic fields are among the most intensely studied number fields. They are of the form $Q(\zeta_n)$, where ζ_n is a primitive n-th root of unity, i.e., a complex number satisfying $\zeta^n = 1$ and $\zeta^m \neq 1$ for all $m < n$. For n being a regular prime, Kummer used cyclotomic fields to prove Fermat's last theorem, which asserts the non-existence of rational nonzero solutions to the equation

$$x^n + y^n = z^n.$$

Local fields are completions of global fields. Ostrowski's theorem asserts that the only completions of Q, a global field, are the local fields Q_p and R. Studying arithmetic questions in global fields may sometimes be done by looking at the corresponding questions locally. This technique is called the local-global principle. For example, the Hasse–Minkowski theorem reduces the problem of finding rational solutions of quadratic equations to solving these equations in R and Q_p, whose solutions can easily be described.

Unlike for local fields, the Galois groups of global fields are not known. Inverse Galois theory studies the (unsolved) problem whether any finite group is the Galois group $Gal(F/Q)$ for some number field F. Class field theory describes the abelian extensions, i.e., ones with abelian Galois group, or equivalently the abelianized Galois groups of global fields. A classical statement, the Kronecker–

Weber theorem, describes the maximal abelian Q^{ab} extension of Q: it is the field

$$Q(\zeta_n, n \geq 2)$$

obtained by adjoining all primitive n-th roots of unity. Kronecker's Jugendtraum asks for a similarly explicit description of F^{ab} of general number fields F. For imaginary quadratic fields, $F = \mathbf{Q}(\sqrt{-d}), d > 0$, the theory of complex multiplication describes F^{ab} using elliptic curves. For general number fields, no such explicit description is known.

Related Notions

In addition to the additional structure that fields may enjoy, fields admit various other related notions. Since in any field $0 \neq 1$, any field has at least two elements. Nonetheless, there is a concept of field with one element, which is suggested to be a limit of the finite fields F_p, as p tends to 1. In addition to division rings, there are various other weaker algebraic structures related to fields such as quasifields, near-fields and semifields.

There are also proper classes with field structure, which are sometimes called Fields, with a capital F. The surreal numbers form a Field containing the reals, and would be a field except for the fact that they are a proper class, not a set. The nimbers, a concept from game theory form a Field.

Division Rings

The hairy ball theorem states that a ball can not be combed. More formally, there is no tangent vector field on the sphere S^2, which is everywhere non-zero.

Dropping one or several axioms in the definition of a field leads to other algebraic structures. As was mentioned above, commutative rings satisfy all axioms of fields, except for multiplicative inverses. Dropping instead the condition that multiplication is commutative leads to the concept of a *division ring* or *skew field*. The only division rings that are finite-dimensional R-vector spaces are R itself, C (which is a field), the quaternions H (in which multiplication is non-commutative), and the octonions O (in which multiplication is neither commutative nor associative). This fact was proved using methods of algebraic topology in 1958 by Michel Kervaire, Raoul Bott, and John Milnor. The non-existence of an odd-dimensional division algebra is more classical. It can be deduced from the hairy ball theorem illustrated in the above.

References

- Besche, Hans Ulrich; Eick, Bettina; O'Brien, E. A. (2001), "The groups of order at most 2000", Electronic Research Announcements of the American Mathematical Society, 7: 1–4, doi:10.1090/S1079-6762-01-00087-7, MR 1826989
- Jahn, H.; Teller, E. (1937), "Stability of Polyatomic Molecules in Degenerate Electronic States. I. Orbital Degeneracy", Proceedings of the Royal Society A, 161 (905): 220–235, Bibcode:1937RSPSA.161..220J, doi:10.1098/rspa.1937.0142
- Michel, Anthony N.; Herget, Charles J. (1993), Applied Algebra and Functional Analysis, New York: Dover Publications, ISBN 978-0-486-67598-5
- Kleiner, Israel (1986), "The evolution of group theory: a brief survey", Mathematics Magazine, 59 (4): 195–215, doi:10.2307/2690312, ISSN 0025-570X, JSTOR 2690312, MR 0863090
- Burris, Stanley N.; Sankappanavar, H. P. (1981), A Course in Universal Algebra, Berlin, New York: Springer-Verlag, ISBN 978-3-540-90578-3
- Israel Kleiner (1996) "The Genesis of the Abstract Ring Concept", American Mathematical Monthly 103: 417–424 doi:10.2307/2974935
- Mac Lane, Saunders (1998), Categories for the Working Mathematician (2nd ed.), Berlin, New York: Springer-Verlag, ISBN 978-0-387-98403-2
- Jacobson, Nathan (1945), "Structure theory of algebraic algebras of bounded degree", Annals of Mathematics, Annals of Mathematics, 46 (4): 695–707, doi:10.2307/1969205, ISSN 0003-486X, JSTOR 1969205
- Taylor, Paul (1999), Practical foundations of mathematics, Cambridge University Press, ISBN 978-0-521-63107-5
- Lam, T. Y. (2001), A First Course in Noncommutative Rings, Graduate Texts in Mathematics, 131 (Second ed.), New York: Springer-Verlag, doi:10.1007/978-1-4419-8616-0, ISBN 0-387-95183-0, MR 1838439
- Devlin, Keith (2000), The Language of Mathematics: Making the Invisible Visible, Owl Books, ISBN 978-0-8050-7254-9
- Lang, Serge (2002), Algebra, Graduate Texts in Mathematics, 211 (Revised third ed.), New York: Springer-Verlag, ISBN 978-0-387-95385-4, MR 1878556
- Cannon, John J. (1969), "Computers in group theory: A survey", Communications of the ACM, 12: 3–12, doi:10.1145/362835.362837, MR 0290613
- Lang, Serge (2005), Undergraduate Algebra (3rd ed.), Berlin, New York: Springer-Verlag, ISBN 978-0-387-22025-3
- Gilmer, R.; Mott, J. (1973). "Associative Rings of Order". Proc. Japan Acad. 49: 795–799. doi:10.3792/pja/1195519146
- Robinson, Derek John Scott (1996), A course in the theory of groups, Berlin, New York: Springer-Verlag, ISBN 978-0-387-94461-6
- Suzuki, Michio (1951), "On the lattice of subgroups of finite groups", Transactions of the American Mathematical Society, 70 (2): 345–371, doi:10.2307/1990375, JSTOR 1990375

3

Universal Algebra

The field of mathematics that studies algebraic structures is known as universal algebra. The study of variety, clone, structure, post's lattice, etc. is fundamental to this field. This chapter closely examines these algebraic structures to provide an extensive understanding of the subject.

Universal Algebra

Universal algebra (sometimes called general algebra) is the field of mathematics that studies algebraic structures themselves, not examples ("models") of algebraic structures. For instance, rather than take particular groups as the object of study, in universal algebra one takes the class of groups as an object of study.

Basic Idea

In universal algebra, an algebra (or algebraic structure) is a set A together with a collection of operations on A. An n-ary operation on A is a function that takes n elements of A and returns a single element of A. Thus, a 0-ary operation (or *nullary operation*) can be represented simply as an element of A, or a *constant*, often denoted by a letter like a. A 1-ary operation (or *unary operation*) is simply a function from A to A, often denoted by a symbol placed in front of its argument, like $\sim x$. A 2-ary operation (or *binary operation*) is often denoted by a symbol placed between its arguments, like $x * y$. Operations of higher or unspecified *arity* are usually denoted by function symbols, with the arguments placed in parentheses and separated by commas, like $f(x,y,z)$ or $f(x_1,...,x_n)$. Some researchers allow infinitary operations, such as $\bigwedge_{\alpha\in J} x_\alpha$ where J is an infinite index set, thus leading into the algebraic theory of complete lattices. One way of talking about an algebra, then, is by referring to it as an algebra of a certain type Ω, where Ω is an ordered sequence of natural numbers representing the arity of the operations of the algebra.

It is also possible to define an algebra via the relations in the algebra instead of the operations. Birkhoff's Theorem states that the two definitions are equivalent, i.e., there is a Galois connection between relational and operational structures. This connection can be easily illustrated on the case of lattices, where the algebraic structure can be given by the operations join and meet or by introducing a partial order relation. The relational point of view is useful in computational problems, in particular for the constraint satisfaction problem (CSP).

Equations

After the operations have been specified, the nature of the algebra is further defined by axioms, which in universal algebra often take the form of identities, or equational laws. An example is the associative axiom for a binary operation, which is given by the equation $x * (y * z) = (x * y) * z$. The axiom is intended to hold for all elements x, y, and z of the set A.

Varieties

A collection of algebraic structures defined by identities is called a variety or equational class. Some authors consider varieties to be the main focus of universal algebra.

Restricting one's study to varieties rules out:

- quantification, including universal quantification ($\forall$), except before an equation, and existential quantification ($\exists$)
  - Predicate logic in particular is ruled out
- All relations except equality, in particular inequalities, both $a \neq b$ and order relations

The study of equational classes can be seen as a special branch of model theory, typically dealing with structures having operations only (i.e. the type can have symbols for functions but not for relations other than equality), and in which the language used to talk about these structures uses equations only.

Not all algebraic structures in a wider sense fall into this scope. For example, ordered groups involve an ordering relation, so would not fall within this scope.

The class of fields is not an equational class because there is no type (or "signature") in which all field laws can be written as equations (inverses of elements are defined for all *non-zero* elements in a field, so inversion cannot be added to the type).

One advantage of this restriction is that the structures studied in universal algebra can be defined in any category that has *finite products*. For example, a topological group is just a group in the category of topological spaces.

Examples

Most of the usual algebraic systems of mathematics are examples of varieties, but not always in an obvious way – the usual definitions often involve quantification or inequalities.

Groups

To see how this works, let's consider the definition of a group. Normally a group is defined in terms of a single binary operation $*$, subject to these axioms:

- Associativity (as in the previous section): $x * (y * z) = (x * y) * z$; formally: $\forall x,y,z.\ x*(y*z)=(x*y)*z$.
- Identity element: There exists an element e such that for each element x, $e * x = x = x * e$; formally: $\exists e\ \forall x.\ e*x=x=x*e$.
- Inverse element: It can easily be seen that the identity element is unique. If this unique

identity element is denoted by e then for each x, there exists an element i such that $x * i = e = i * x$; formally: $\forall x\, \exists i.\, x*i=e=i*x$.

(Some authors also use an axiom called "closure", stating that $x * y$ belongs to the set A whenever x and y do. But from a universal algebraist's point of view, that is already implied by calling $*$ a binary operation.)

This definition of a group is problematic from the point of view of universal algebra. The reason is that the axioms of the identity element and inversion are not stated purely in terms of equational laws but also have clauses involving the phrase "there exists ... such that ...". This is inconvenient; the list of group properties can be simplified to universally quantified equations by adding a nullary operation e and a unary operation ~ in addition to the binary operation $*$. Then list the axioms for these three operations as follows:

- Associativity: $x * (y * z) = (x * y) * z$.
- Identity element: $e * x = x = x * e$; formally: $\forall x.\, e*x=x=x*e$.
- Inverse element: $x * (\sim x) = e = (\sim x) * x$ formally: $\forall x.\, x*\sim x=e=\sim x*x$.

(Of course, we usually write "x^{-1}" instead of "$\sim x$", which shows that the notation for operations of low arity is not *always* as given in the second paragraph.)

What has changed is that in the usual definition there are:

- a single binary operation (signature (2))
- 1 equational law (associativity)
- 2 quantified laws (identity and inverse)

...while in the universal algebra definition there are

- 3 operations: one binary, one unary, and one nullary (signature (2,1,0))
- 3 equational laws (associativity, identity, and inverse)
- no quantified laws (except for outermost universal quantifiers which are allowed in varieties)

It is important to check that this really does capture the definition of a group. The reason that it might not is that specifying one of these universal groups might give more information than specifying one of the usual kind of group. After all, nothing in the usual definition said that the identity element e was *unique*; if there is another identity element e', then it is ambiguous which one should be the value of the nullary operator e. Proving that it is unique is a common beginning exercise in classical group theory textbooks. The same thing is true of inverse elements. So, the universal algebraist's definition of a group is equivalent to the usual definition.

At first glance this is simply a technical difference, replacing quantified laws with equational laws. However, it has immediate practical consequences – when defining a group object in category theory, where the object in question may not be a set, one must use equational laws (which make sense in general categories), and cannot use quantified laws (which do not make sense, as objects in general categories do not have elements). Further, the perspective of universal algebra insists not only that the inverse and identity exist, but that they be maps in the category. The basic example is a topological group – not only must the inverse exist element-wise, but the inverse map must be continuous (some authors also require the identity map to be a closed inclusion, hence cofibration, again referring to properties of the map).

Other Examples

Most algebraic structures are examples of universal algebras.

- Rings, semigroups, quasigroups, groupoids, magmas, loops, and others.
- Vector spaces or modules over a fixed field or rings are universal algebras. The operations, in this case, are addition and, for each element of the field or ring, scalar multiplication by this element.

Examples of relational algebras include semilattices, lattices or Boolean algebras.

Basic Constructions

We assume that the type, Ω, has been fixed. Then there are three basic constructions in universal algebra: homomorphic image, subalgebra, and product.

A homomorphism between two algebras A and B is a function $h: A \rightarrow B$ from the set A to the set B such that, for every operation f_A of A and corresponding f_B of B (of arity, say, n), $h(f_A(x_1,...,x_n)) = f_B(h(x_1),...,h(x_n))$. (Sometimes the subscripts on f are taken off when it is clear from context which algebra the function is from.) For example, if e is a constant (nullary operation), then $h(e_A) = e_B$. If $\sim$ is a unary operation, then $h(\sim x) = \sim h(x)$. If $*$ is a binary operation, then $h(x * y) = h(x) * h(y)$. And so on. A few of the things that can be done with homomorphisms, as well as definitions of certain special kinds of homomorphisms, are listed under the entry Homomorphism. In particular, we can take the homomorphic image of an algebra, $h(A)$.

A subalgebra of A is a subset of A that is closed under all the operations of A. A product of some set of algebraic structures is the cartesian product of the sets with the operations defined coordinatewise.

Some Basic Theorems

- The isomorphism theorems, which encompass the isomorphism theorems of groups, rings, modules, etc.
- Birkhoff's HSP Theorem, which states that a class of algebras is a variety if and only if it is closed under homomorphic images, subalgebras, and arbitrary direct products.

Motivations and Applications

In addition to its unifying approach, universal algebra also gives deep theorems and important examples and counterexamples. It provides a useful framework for those who intend to start the study of new classes of algebras. It can enable the use of methods invented for some particular classes of algebras to other classes of algebras, by recasting the methods in terms of universal algebra (if possible), and then interpreting these as applied to other classes. It has also provided conceptual clarification; as J.D.H. Smith puts it, *"What looks messy and complicated in a particular framework may turn out to be simple and obvious in the proper general one."*

In particular, universal algebra can be applied to the study of monoids, rings, and lattices. Before universal algebra came along, many theorems (most notably the isomorphism theorems) were proved separately in all of these classes, but with universal algebra, they can be proven once and for all for every kind of algebraic system.

The 1956 paper by Higgins referenced below has been well followed up for its framework for a range of particular algebraic systems, while his 1963 paper is notable for its discussion of algebras with operations which are only partially defined, typical examples for this being categories and groupoids. This leads on to the subject of higher-dimensional algebra which can be defined as the study of algebraic theories with partial operations whose domains are defined under geometric conditions. Notable examples of these are various forms of higher-dimensional categories and groupoids.

Constraint Satisfaction Problem

Universal algebra provides a natural language for the constraint satisfaction problem (CSP). CSP refers to an important class of computational problems where given a relational algebra A and an existential sentence φ over this algebra, the question is to find out whether φ can be satisfied in A. The algebra A is often fixed so that CSP_A refers to the problem whose instance is only the existential sentence φ.

It is proved that every computational problem can be formulated as CSP_A for some algebra A.

For example, the n-coloring problem can be stated as CSP of the algebra $(\{0,1,\ldots,n-1\},\neq)$, i.e., an algebra with n elements and a single relation inequality.

The dichotomy conjecture (proved in April 2017) states that if A is a finite algebra, then CSP_A is either P or NP-complete.

Generalizations

Universal algebra has also been studied using the techniques of category theory. In this approach, instead of writing a list of operations and equations obeyed by those operations, one can describe an algebraic structure using categories of a special sort, known as Lawvere theories or more generally algebraic theories. Alternatively, one can describe algebraic structures using monads. The two approaches are closely related, with each having their own advantages. In particularly, every Lawvere theory gives a monad on the category of sets, while any 'finitary' monad on the category of sets arises from a Lawvere theory. However, a monad describes algebraic structures within one particular category (for example the category of sets), while algebraic theories describe structure within any of a large class of categories (namely those having finite products).

A more recent development in category theory is operad theory – an operad is a set of operations, similar to a universal algebra, but restricted in that equations are only allowed between expressions with the variables, with no duplication or omission of variables allowed. Thus, rings can be described as the so-called algebras of some operad, but not groups, since the law $gg^{-1} = 1$ duplicates the variable g on the left side and omits it on the right side. At first this may seem to be a troublesome restriction, but the payoff is that operads have certain advantages: for example, one can hybridize the concepts of ring and vector space to obtain the concept of associative algebra, but one cannot form a similar hybrid of the concepts of group and vector space.

Another development is partial algebra where the operators can be partial functions. Certain partial functions can also be handled by a generalization of Lawvere theories known as essentially algebraic theories.

Another generalization of universal algebra is model theory, which is sometimes described as "universal algebra + logic".

Basis (Universal Algebra)

In universal algebra a basis is a structure inside of some (universal) algebras, which are called free algebras. It generates all algebra elements from its own elements by the algebra operations in an independent manner. It also represents the endomorphisms of an algebra by certain indexings of algebra elements, which can correspond to the usual matrices when the free algebra is a vector space.

Definitions

The basis (or reference frame) of a (universal) algebra is a function b that takes some algebra elements as values $b(i)$ and satisfies either one of the following two equivalent conditions. Here, the set of all $b(i)$ is called basis set, whereas several authors call it the "basis". The set I of its arguments i is called dimension set. Any function, with all its arguments in the whole I, that takes algebra elements as values (even outside the basis set) will be denoted by m. Then, b will be an m.

Outer Condition

This condition will define bases by the set L of the I-ary elementary functions of the algebra, which are certain functions ℓ that take every m as argument to get some algebra element as value $\ell(m)$. In fact, they consist of all the projections p_i with i in I, which are the functions such that $p_i(m) = m(i)$ for each m, and of all functions that rise from them by repeated "multiple compositions" with operations of the algebra.

(When an algebra operation has a single algebra element as argument, the value of such a composed function is the one that the operation takes from the value of a single previously computed I-ary function as in composition. When it does not, such compositions require that many (or none for a nullary operation) I-ary functions are evaluated before the algebra operation: one for each possible algebra element in that argument. In case I and the numbers of elements in the arguments, or "arity", of the operations are finite, this is the finitary multiple composition.)

Then, according to the *outer condition* a basis has to *generate* the algebra (namely when ℓ ranges over the whole L, $\ell(b)$ gets every algebra element) and must be *independent* (namely whenever any two I -ary elementary functions coincide at b, they will do everywhere: $\ell'(b) = \ell''(b)$ implies $\ell' = \ell''$). This is the same as to require that there exists a *single* function χ that takes every algebra element as argument to get an I -ary elementary function as value and satisfies $\chi(\ell(b)) = \ell$ for all ℓ in L.

Inner Condition

This other condition will define bases by the set E of the endomorphisms of the algebra, which are the homomorphisms from the algebra into itself, through its analytic representation ϱ by a basis. The latter is a function that takes every endomorphism e as argument to get a function m as value: $\varrho(e) = m$, where this m is the "sample" of the values of e at b, namely $m(i) = [\varrho(e)]_i = e(b(i))$ for all i in the dimension set.

Then, according to the *inner condition* b is a basis, when ϱ is a bijection from E onto the set of all m, namely for each m there is one and only one endomorphism e such that $m = \varrho(e)$. This is the same as to require that there exists an extension function, namely a function η that takes every (sample) m as argument to extend it onto an endomorphism $\eta(m)$ such that $\varrho(\eta(m)) = m$.

The link between these two conditions is given by the identity $[\chi(a)]_m = [\eta(m)]_a$, which holds for all m and all algebra elements a. Several other conditions that characterize bases for universal algebras are omitted.

As the next example will show, present bases are a generalization of the bases of vector spaces. Then, the name "reference frame" can well replace "basis". Yet, contrary to the vector space case, a universal algebra might lack bases and, when it has them, their dimension sets might have different finite positive cardinalities.

Examples

Vector Space Algebras

In the universal algebra corresponding to a vector space with positive dimension the bases essentially are the ordered bases of this vector space. Yet, this will come after several details.

When the vector space is finite-dimensional, for instance $I = \{0,1,\ldots n-1\}$ with $n > 0$, the functions ℓ in the set L of the *outer condition* exactly are the ones that provide the spanning and linear independence properties with linear combinations $\ell(b) = c_0 b_0 + c_1 b_1 + \ldots c_{n-1} b_{n-1}$ and present generator property becomes the spanning one. On the contrary, linear independence is a mere instance of present independence, which becomes equivalent to it in such vector spaces. (Also, several other generalizations of linear independence for universal algebras do not imply present independence.)

The functions m for the *inner condition* correspond to the square arrays of field numbers (namely, usual vector-space square matrices) that serve to build the endomorphisms of vector spaces (namely, linear maps into themselves). Then, the *inner condition* requires a bijection property from endomorphisms also to arrays. In fact, each column of such an array represents a vector $m(i)$ as its n-tuple of coordinates with respect to the basis b. For instance, when the vectors are n-tuples of numbers from the underlying field and b is the Kronecker basis, m is such an array *seen by columns*, ϱ is the sample of such a linear map at the reference vectors and η extends this sample to this map as below.

$$\begin{pmatrix} 0 & -1 & 2 \\ -2 & 3 & 1 \\ 1 & 0 & 2 \end{pmatrix} \overset{\eta}{\underset{\varrho}{\rightleftarrows}} \begin{cases} x_0' = & & -x_1 & + & 2x_2 \\ x_1' = & -2x_0 & +3x_1 & + & x_2 \\ x_2' = & x_0 & & + & 2x_2 \end{cases}$$

When the vector space is not finite-dimensional, further distinctions are needed. In fact, though the functions ℓ formally have an infinity of vectors in every argument, the linear combinations they evaluate never require infinitely many addenda $c_i m(i)$ and each ℓ determines a finite subset J of I that contains all required i. Then, every value $\ell(m)$ equals an $\ell'(m')$, where m' is the restriction of m to J and ℓ' is the J-ary elementary function corresponding to ℓ. When the ℓ' replace the ℓ, both the linear independence and spanning properties for infinite basis sets follow from present *outer condition* and conversely.

Therefore, as far as vector spaces of a positive dimension are concerned, the only difference between present bases for universal algebras and the ordered bases of vector spaces is that here no order on I is required. Still it is allowed, in case it serves some purpose.

When the space is zero-dimensional, its ordered basis is empty. Then, being the empty function, it is a present basis. Yet, since this space only contains the null vector and its only endomorphism is the identity, any function b from any set I (even a nonempty one) to this singleton space work as a present basis. This is not so strange from the point of view of Universal Algebra, where singleton algebras, which are called "trivial", enjoy a lot of other seeming strange properties.

Word Monoid

Let $I = \{\mathrm{a}, \mathrm{b}, \mathrm{c}, \ldots\}$ be an "alphabet", namely a (usually finite) set of objects called "letters". Let W denote the corresponding set of words or "strings", which will be denoted as in strings, namely either by writing their letters in sequence or by ϵ in case of the empty word (Formal Language notation). Accordingly, the juxtaposition vw will denote the concatenation of two words v and w, namely the word that begins with v and is followed by w.

Concatenation is a binary operation on W that together with the empty word ϵ defines a free monoid, the monoid of the words on I, which is one of the simplest universal algebras. Then, the *inner condition* will immediately prove that one of its bases is the function b that makes a single-letter word i of each letter i, $b(\mathrm{i}) = i$.

(Depending on the set-theoretical implementation of sequences, b may not be an identity function, namely i may not be i, rather an object like $\{(\varnothing, \mathrm{i})\}$, namely a singleton function, or a pair like $(\varnothing, \mathrm{i})$ or $(\mathrm{i}, \varnothing)$.)

In fact, in the theory of DoL systems such $m = \varrho(e)$ are the tables of "productions", which such systems use to define the simultaneous substitutions of every by a single word $w = m(\mathrm{i})$ in any word u in W: if $u = i_0 i_1 \cdots i_k$, then $e(u) = m(\mathrm{i}_0) m(\mathrm{i}_1) \cdots m(\mathrm{i}_k)$. Then, b satisfies the *inner condition*, since the function ϱ is the well-known bijection that identifies every word endomorphism with any such table. (The repeated applications of such an endomorphism starting from a given "seed" word are able to model many growth processes, where words and concatenation serve to build fairly heterogeneous structures as in L-system, not just "sequences".)

Variety

In the mathematical subject of universal algebra, a variety of algebras is the class of all algebraic structures of a given signature satisfying a given set of identities. For example, the groups form a variety of algebras, as do the abelian groups, the rings, the monoids etc. According to Birkhoff's theorem, a class of algebraic structures of the same signature is a variety if and only if it is closed under the taking of homomorphic images, subalgebras and (direct) products. In the context of category theory, a variety of algebras, together with its homomorphisms, forms a category; these are usually called finitary algebraic categories.

A covariety is the class of all coalgebraic structures of a given signature.

Terminology

A variety of algebras should not be confused with an algebraic variety. Intuitively, a variety of algebras is a *collection of algebras*, defined by a set of equations that must be identically satisfied in each algebra, while an algebraic variety is a *collection of elements from a single algebra,* defined by a set of equations that these elements must satisfy. The two are named alike by analogy, but they are formally quite distinct and their theories have little in common.

The term "variety of algebras" refers to algebras in the general sense of universal algebra; there is also a more specific sense of algebra, namely as algebra over a field, i.e. a vector space equipped with a multiplication.

Examples

The class of all semigroups forms a variety of algebras of signature (2), meaning that a semigroup has a single binary operation. A sufficient defining equation is the associative law:

$$x(yz) = (xy)z.$$

The class of groups forms a variety of algebras of signature (2,0,1), the three operations being respectively *multiplication* (binary), *identity* (nullary, a constant) and *inversion* (unary). The familiar axioms of associativity, inverse and identity form one suitable set of identities:

$$x(yz) = (xy)z$$

$$1x = x1 = x$$

$$xx^{-1} = x^{-1}x = 1.$$

The class of rings also forms a variety of algebras. The signature here is (2,2,0,0,1) (two binary operations, two constants, and one unary operation).

If we fix a specific ring R, we can consider the class of left R-modules. To express the scalar multiplication with elements from R, we need one unary operation for each element of R. If the ring is infinite, we will thus have infinitely many operations, which is allowed by the definition of an algebraic structure in universal algebra. We will then also need infinitely many identities to express the module axioms, which is allowed by the definition of a variety of algebras. So the left R-modules do form a variety of algebras.

The fields do *not* form a variety of algebras; the requirement that all non-zero elements be invertible cannot be expressed as a universally satisfied identity.

The cancellative semigroup also don't form a variety of algebras, since the cancellation property is an implication and not an equation.

Birkhoff's Theorem

Given a class of algebraic structures of the same signature, we can define the notions of homomorphism, subalgebra, and product. Garrett Birkhoff proved that a class of algebraic structures of the same signature is a variety if and only if it is closed under the taking of homomorphic images, subalgebras and arbitrary products. This is a result of fundamental importance to universal algebra and known as Birkhoff's theorem or as the HSP theorem. H, S, and P stand, respectively, for the operations of homomorphism, subalgebra, and product.

It is easy to see that the class of algebras satisfying some set of identities will be closed under the HSP operations. Proving the converse—classes of algebras closed under the HSP operations must be equational—is harder.

Using Birkhoff's theorem, we can for example verify the claim made above, that the field axioms are not expressable by any possible set of identities: the product of fields is not a field, so fields don't form a variety.

Subvarieties

A subvariety of a variety of algebras V is a subclass of V that has the same signature as V and is itself a variety, i.e., is defined by a set of identities.

Notice that although every group becomes a semigroup when the identity as a constant is omitted (and/or the inverse operation is omitted), the class of groups does *not* form a subvariety of the variety of semigroups because the signatures are different. Similarly, the class of semigroups which are groups is not a subvariety of the variety of semigroups. Indeed, the class of monoids which are groups contains $\langle \mathbb{Z}, + \rangle$ and does not not contain its subalgebra (more precesily, submonoid) $\langle \mathbb{N}, + \rangle$.

On the other hand the class of abelian groups is a subvariety of the variety of groups because it consists of those groups satisfying $xy = yx$, with no change of signature. The finitely generated abelian groups do not form a subvariety, since by Birkhoff's theorem they don't form a variety, as an arbitrary product of finitely generated abelian groups is not finitely generated.

Viewing a variety V and its homomorphisms as a category, a subvariety U of V is a full subcategory of V, meaning that for any objects a, b in U, the homomorphisms from a to b in U are exactly those from a to b in V.

Free Objects

Suppose V is a non-trivial variety of algebras, i.e. V contains algebras with more than one element. One can show that for every set S, the variety V contains a *free algebra* F_S *on* S. This means that there is an injective set map $i : S \to F_S$ which satisfies the following universal property: given any algebra A in V and any map $k : S \to A$, there exists a unique V-homomorphism $f : F_S \to A$ such that $fi = k$.

This generalizes the notions of free group, free abelian group, free algebra, free module etc. It has the consequence that every algebra in a variety is a homomorphic image of a free algebra.

Category Theory

If V is a finitary algebraic category (i.e. the category of a variety of algebras, with homomorphisms as morphisms) then the forgetful functor

$$G : V \to \mathbf{Set}$$

has a left adjoint $F : \mathbf{Set} \to V$, namely the functor that assigns to each set the free algebra on that set. This adjunction is *strictly monadic*, in that the category V is isomorphic to the Eilenberg–Moore category $\mathbf{Set}^T$ for the monad $T = GF$.

The monad $T : \mathbf{Set} \to \mathbf{Set}$ is thus enough to recover the finitary algebraic category, which allows the following generalization. One says a category is an algebraic category if it is monadic over $\mathbf{Set}$. This is a more general notion than "finitary algebraic category" because it admits such categories as CABA (complete atomic Boolean algebras) and CSLat (complete semilattices) whose signatures include infinitary operations. In those two cases the signature is large, meaning that it forms not a set but a proper class, because its operations are of unbounded arity. The algebraic category of sigma algebras also has infinitary operations, but their arity is countable whence its signature is small (forms a set).

Every finitary algebraic category is a locally presentable category.

Pseudovariety of Finite Algebras

Since varieties are closed under arbitrary direct products, all non-trivial varieties contain infinite algebras. Attempts have been made to develop a finitary analogue of the theory of varieties. This led, e.g., to the notion of variety of finite semigroups. This kind of variety uses only finitary products. However, it uses a more general kind of identities.

A pseudovariety is usually defined to be a class of algebras of a given signature, closed under the taking of homomorphic images, subalgebras and finitary direct products. Not every author assumes that all algebras of a pseudovariety are finite; if this is the case, one sometimes talks of a variety of finite algebras. For pseudovarieties, there is no general finitary counterpart to Birkhoff's theorem, but in many cases the introduction of a more complex notion of equations allows similar results to be derived.

Pseudovarieties are of particular importance in the study of finite semigroups and hence in formal language theory. Eilenberg's theorem, often referred to as the *variety theorem*, describes a natural correspondence between varieties of regular languages and pseudovarieties of finite semigroups.

Clone

In universal algebra, a clone is a set C of finitary operations on a set A such that

- C contains all the projections $\pi_k^n : A^n \to A$, *defined by* $\pi_k^n(x_1, \ldots, x_n) = x_k$,
- C is closed under (finitary multiple) composition (or "superposition"): if $f, g_1, \ldots, g_m$ are members of C such that if f is m-ary, and g_j is n-ary for every j, then the n-ary operation $h(x_1, \ldots, x_n) := f(g_1(x_1, \ldots, x_n), \ldots, g_m(x_1, \ldots, x_n))$ is in C.

The question whether clones should contain nullary operations or not is not treated uniformly in the literature. The classical approach as evidenced by the standard monographs on clone theory considers clones only containing at least unary operations. However, with only minor modifications (related to the empty invariant relation) most of the usual theory can be lifted to clones allowing nullary operations. The more general concept includes all clones without nullary operations as subclones of the clone of all at least unary operations and is in accordance with the custom to allow nullary terms and nullary term operations in universal algebra. Typically, publications studying clones as abstract clones, e.g. in the category theoretic setting of Lawvere's algebraic theories, will include nullary operations.

Given an algebra in a signature σ, the set of operations on its carrier definable by a σ-term (the *term functions*) is a clone. Conversely, every clone can be realized as the clone of term functions in a suitable algebra by simply taking the clone itself as source for the signature σ so that the algebra has the whole clone as its fundamental operations.

If A and B are algebras with the same carrier such that every basic function of A is a term function in B and vice versa, then A and B have the same clone. For this reason, modern universal algebra often treats clones as a representation of algebras which abstracts from their signature.

There is only one clone on the one-element set (there are two if nullary operations are considered). The lattice of clones on a two-element set is countable, and has been completely described by Emil Post. Clones on larger sets do not admit a simple classification; there are continuum clones on a finite set of size at least three, and 2^{2^κ} clones on an infinite set of cardinality κ.

Abstract Clones

Philip Hall introduced the concept of *abstract clone*. An abstract clone is different from a concrete clone in that the set A is not given. Formally, an abstract clone comprises

- a set C_n for each natural number n,
- elements $\pi_{k,n}$ in C_n for all $k \le n$, and
- a family of functions $*: C_m \times (C_n)^m \to C_n$ for all m and n

such that

- $c * (\pi_{1,n}, \ldots, \pi_{n,n}) = c$

- $\pi_{k,m} * (c_1,...,c_m) = c_k$
- $c * (d_1 * (e_1, ..., e_n), ..., d_m * (e_1, ..., e_n)) = (c * (d_1,..., d_m)) * (e_1,...,e_n)$.

Any concrete clone determines an abstract clone in the obvious manner.

Any algebraic theory determines an abstract clone where C_n is the set of terms in n variables, $\pi_{k,n}$ are variables, and $*$ is substitution. Two theories determine isomorphic clones if and only if the corresponding categories of algebras are isomorphic. Conversely every abstract clone determines an algebraic theory with an n-ary operation for each element of C_n. This gives a bijective correspondence between abstract clones and algebraic theories.

Every abstract clone C induces a Lawvere theory in which the morphisms $m \to n$ are elements of $(C_m)^n$. This induces a bijective correspondence between Lawvere theories and abstract clones.

BCK Algebra

In mathematics, BCI and BCK algebras are algebraic structures, introduced by Y. Imai, K. Iséki and S. Tanaka in 1966, that describe fragments of the propositional calculus involving implication known as BCI and BCK logics.

Definition

BCI Algebra

An algebra $(X;*,0)$ of type $(2,0)$ is called a *BCI-algebra* if, for any $x, y, z \in X$, it satisfies the following conditions. (Informally, we may read 0 as "truth" and $x*y$ as "y implies x".)

BCI-1

$$((x*y)*(x*z))*(z*y)=0$$

BCI-2

$$(x*(x*y))*y=0$$

BCI-3

$$x*x=0$$

BCI-4

$$x*y=0 \wedge y*x=0 \Rightarrow x=y$$

BCI-5

$$x*0=0 \Rightarrow x=0$$

BCK algebra

A BCI-algebra $(X;*,0)$ is called a *BCK-algebra* if it satisfies the following condition:

BCK-1

$$\forall x \in X : 0 * x = 0.$$

A partial order can then be defined as $x \le y$ iff $x * y = 0$.

A BCK-algebra is said to be *commutative* if it satisfies:

$$x * (x * y) = y * (y * x)$$

In a commutative BCK-algebra $x * (x * y) = x \wedge y$ is the greatest lower bound of x and y under the partial order $\le$.

A BCK-algebra is said to be bounded if it has a largest element, usually denoted by 1. In a bounded commutative BCK-algebra the least upper bound of two elements satisfies $x \vee y = 1 * ((1 * x) \wedge (1 * y))$; that makes it a distributive lattice.

Examples

Every abelian group is a BCI-algebra, with * defined as group subtraction and 0 defined as the group identity.

The subsets of a set form a BCK-algebra, where A*B is the difference A\B (the elements in A but not in B), and 0 is the empty set.

A Boolean algebra is a BCK algebra if $A*B$ is defined to be $A \wedge \neg B$ (A does not imply B).

The bounded commutative BCK-algebras are precisely the MV-algebras.

Post's Lattice

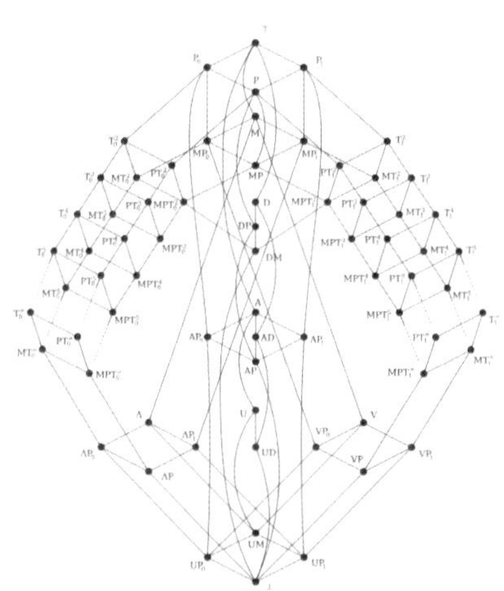

Hasse diagram of Post's lattice.

In logic and universal algebra, Post's lattice denotes the lattice of all clones on a two-element set {0, 1}, ordered by inclusion. It is named for Emil Post, who published a complete description of the lattice in 1941. The relative simplicity of Post's lattice is in stark contrast to the lattice of clones on a three-element (or larger) set, which has the cardinality of the continuum, and a complicated inner structure. A modern exposition of Post's result can be found in Lau (2006).

Basic Concepts

A Boolean function, or logical connective, is an n-ary operation f: $2^n \to 2$ for some $n \geq 1$, where 2 denotes the two-element set {0, 1}. Particular Boolean functions are the projections

$$\pi_k^n(x_1,\ldots,x_n) = x_k,$$

and given an m-ary function f, and n-ary functions $g_1, \ldots, g_m$, we can construct another n-ary function

$$h(x_1,\ldots,x_n) = f(g_1(x_1,\ldots,x_n),\ldots,g_m(x_1,\ldots,x_n)),$$

called their composition. A set of functions closed under composition, and containing all projections, is called a clone.

Let B be a set of connectives. The functions which can be defined by a formula using propositional variables and connectives from B form a clone $[B]$, indeed it is the smallest clone which includes B. We call $[B]$ the clone *generated* by B, and say that B is the *basis* of $[B]$. For example, $[\neg, \wedge]$ are all Boolean functions, and $[0, 1, \wedge, \vee]$ are the monotone functions.

We use the operations $\neg$, Np, (negation), $\wedge$, Kpq, (conjunction or meet), $\vee$, Apq, (disjunction or join), $\to$, Cpq, (implication), $\leftrightarrow$, Epq, (biconditional), +, Jpq (exclusive disjunction or Boolean ring addition), $\nrightarrow$, Lpq, (nonimplication), ?: (the ternary conditional operator) and the constant unary functions 0 and 1. Moreover, we need the threshold functions.

$$\text{th}_k^n(x_1,\ldots,x_n) = \begin{cases} 1 & \text{if } \left|\{i \mid x_i = 1\}\right| \geq k, \\ 0 & \text{otherwise.} \end{cases}$$

For example, th_1^n is the large disjunction of all the variables x_i, and th_n^n is the large conjunction. Of particular importance is the majority function

$$\text{maj} = \text{th}_2^3 = (x \wedge y) \vee (x \wedge z) \vee (y \wedge z).$$

We denote elements of $\mathbf{2}^n$ (i.e., truth-assignments) as vectors: a = $(a_1, \ldots, a_n)$. The set $\mathbf{2}^n$ carries a natural product Boolean algebra structure. That is, ordering, meets, joins, and other operations on n-ary truth assignments are defined pointwise:

$$(a_1,\ldots,a_n) \leq (b_1,\ldots,b_n) \Leftrightarrow a_i \leq b_i \text{ for } i = 1,\ldots,n,$$

$$(a_1,\ldots,a_n) \wedge (b_1,\ldots,b_n) = (a_1 \wedge b_1,\ldots,a_n \wedge b_n).$$

Naming of Clones

Intersection of an arbitrary number of clones is again a clone. It is convenient to denote intersection of clones by simple juxtaposition, i.e., the clone $C_1 \cap C_2 \cap ... \cap C_k$ is denoted by $C_1C_2...C_k$. Some special clones are introduced below:

- M is the set of monotone functions: $f(a) \le f(b)$ for every $a \le b$.
- D is the set of self-dual functions: $\neg f(a) = f(\neg a)$.
- A is the set of affine functions: the functions satisfying

  $$f(a_1,\ldots,a_{i-1},c,a_{i+1},\ldots,a_n) = f(a_1,\ldots,d,a_{i+1},\ldots)$$
  $$\Rightarrow f(b_1,\ldots,c,b_{i+1},\ldots) = f(b_1,\ldots,d,b_{i+1},\ldots)$$

  for every $i \le n$, a, b $\in \mathbf{2}^n$, and c, $d \in 2$. Equivalently, the functions expressible as $f(x_1, ..., x_n) = a_0 + a_1x_1 + ... + a_nx_n$ for some a_0, a.
- U is the set of *essentially unary* functions, i.e., functions which depend on at most one input variable: there exists an $i = 1, ..., n$ such that $f(a) = f(b)$ whenever $a_i = b_i$.
- Λ is the set of *conjunctive* functions: $f(\mathbf{a} \wedge \mathbf{b}) = f(\mathbf{a}) \wedge f(\mathbf{b})$. The clone Λ consists of the conjunctions $f(x_1,\ldots,x_n) = \bigwedge_{i\in I} x_i$ for all subsets I of {1, ..., n} (including the empty conjunction, i.e., the constant 1), and the constant 0.
- V is the set of *disjunctive* functions: $f(\mathbf{a} \vee \mathbf{b}) = f(\mathbf{a}) \vee f(\mathbf{b})$. Equivalently, V consists of the disjunctions $f(x_1,\ldots,x_n) = \bigvee_{i\in I} x_i$ for all subsets I of {1, ..., n} (including the empty disjunction 0), and the constant 1.
- For any $k \ge 1$, T_0^k is the set of functions f such that

  $$\mathbf{a}^1 \wedge \cdots \wedge \mathbf{a}^k = \mathbf{0} \Rightarrow f(\mathbf{a}^1) \wedge \cdots \wedge f(\mathbf{a}^k) = 0.$$

  Moreover, $\mathrm{T}_0^\infty = \bigcap_{k=1}^{\infty} \mathrm{T}_0^k$ is the set of functions bounded above by a variable: there exists $i = 1$, ..., n such that $f(a) \le a_i$ for all a.

  As a special case, $P_0 = T_1^1$ is the set of *0-preserving* functions: $f(0) = 0$. Furthermore, ⊤ can be considered T_0^0 when one takes the empty meet into account.
- For any $k \ge 1$, T_1^k is the set of functions f such that

  $$\mathbf{a}^1 \vee \cdots \vee \mathbf{a}^k = \mathbf{1} \Rightarrow f(\mathbf{a}^1) \vee \cdots \vee f(\mathbf{a}^k) = 1,$$

  and $\mathrm{T}_1^\infty = \bigcap_{k=1}^{\infty} \mathrm{T}_1^k$ is the set of functions bounded below by a variable: there exists $i = 1, ..., n$ such that $f(a) \ge a_i$ for all a.

  The special case $P_1 = T_1^1$ consists of the *1-preserving* functions: $f(1) = 1$. Furthermore, ⊤ can be considered T_1^0 when one takes the empty join into account.
- The largest clone of all functions is denoted ⊤, the smallest clone (which contains only projections) is denoted ⊥, and $P = P_0P_1$ is the clone of *constant-preserving* functions.

Applications

The complete classification of Boolean clones given by Post helps to resolve various questions about classes of Boolean functions. For example:

- An inspection of the lattice shows that the maximal clones different from T (often called Post's classes) are M, D, A, P_0, P_1, and every proper subclone of T is contained in one of them. As a set *B* of connectives is functionally complete if and only if it generates T, we obtain the following characterization: *B* is functionally complete iff it is not included in one of the five Post's classes.

- The satisfiability problem for Boolean formulas is NP-complete by Cook's theorem. Consider a restricted version of the problem: for a fixed finite set *B* of connectives, let *B*-SAT be the algorithmic problem of checking whether a given *B*-formula is satisfiable. Lewis used the description of Post's lattice to show that *B*-SAT is NP-complete if the function ↛ can be generated from $B\left(i.e., [B] \supseteq T_0^{\infty}\right)$, and in all the other cases *B*-SAT is polynomial-time decidable.

Variants

Post originally did not work with the modern definition of clones, but with the so-called *iterative systems*, which are sets of operations closed under substitution

$$h(x_1,\ldots,x_{n+m-1}) = f(x_1,\ldots,x_{n-1},g(x_n,\ldots,x_{n+m-1})),$$

as well as permutation and identification of variables. The main difference is that iterative systems do not necessarily contain all projections. Every clone is an iterative system, and there are 20 non-empty iterative systems which are not clones. (Post also excluded the empty iterative system from the classification, hence his diagram has no least element and fails to be a lattice.) As another alternative, some authors work with the notion of a *closed class*, which is an iterative system closed under introduction of dummy variables. There are four closed classes which are not clones: the empty set, the set of constant 0 functions, the set of constant 1 functions, and the set of all constant functions.

Composition alone does not allow to generate a nullary function from the corresponding unary constant function, this is the technical reason why nullary functions are excluded from clones in Post's classification. If we lift the restriction, we get more clones. Namely, each clone *C* in Post's lattice which contains at least one constant function corresponds to two clones under the less restrictive definition: *C*, and *C* together with all nullary functions whose unary versions are in *C*.

Structure

In universal algebra and in model theory, a structure consists of a set along with a collection of finitary operations and relations that are defined on it.

Universal algebra studies structures that generalize the algebraic structures such as groups, rings, fields and vector spaces. The term universal algebra is used for structures with no relation symbols.

Model theory has a different scope that encompasses more arbitrary theories, including foundational structures such as models of set theory. From the model-theoretic point of view, structures are the objects used to define the semantics of first-order logic. For a given theory in model theory, a structure is called a model, if it satisfies the defining axioms of that theory, although it is sometimes disambiguated as a *semantic model* when one discusses the notion in the more general setting of mathematical models. Logicians sometimes refer to structures as interpretations.

In database theory, structures with no functions are studied as models for relational databases, in the form of relational models.

Definition

Formally, a structure can be defined as a triple $\mathcal{A} = (A, \sigma, I)$ consisting of a domain *A*, a signature σ, and an interpretation function *I* that indicates how the signature is to be interpreted on the domain. To indicate that a structure has a particular signature σ one can refer to it as a σ-structure.

Domain

The domain of a structure is an arbitrary set; it is also called the underlying set of the structure, its carrier (especially in universal algebra), or its universe (especially in model theory). In classical first-order logic, the definition of a structure prohibits the empty domain.

Sometimes the notation $\text{dom}(\mathcal{A})$ or $|\mathcal{A}|$ is used for the domain of $\mathcal{A}$, but often no notational distinction is made between a structure and its domain. (I.e. the same symbol $\mathcal{A}$ refers both to the structure and its domain.)

Signature

The signature $\sigma = (S, \text{ar})$ of a structure consists of a set S of function symbols and relation symbols along with a function $\text{ar}: S \to \mathbb{N}_0$ that ascribes to each symbol *s* a natural number $n = \text{ar}(s)$ which is called the arity of *s* because it is the arity of the interpretation of *s*.

Since the signatures that arise in algebra often contain only function symbols, a signature with no relation symbols is called an algebraic signature. A structure with such a signature is also called an algebra; this should not be confused with the notion of an algebra over a field.

Interpretation Function

The interpretation function I of $\mathcal{A}$ assigns functions and relations to the symbols of the signature. Each function symbol *f* of arity *n* is assigned an *n*-ary function $f^{\mathcal{A}} = I(f)$ on the domain. Each relation symbol *R* of arity *n* is assigned an *n*-ary relation $R^{\mathcal{A}} = I(R) \subseteq A^{\text{ar(R)}}$ on the domain. A nullary function symbol c is called a constant symbol, because its interpretation *I(c)* can be identified with a constant element of the domain.

When a structure (and hence an interpretation function) is given by context, no notational distinction is made between a symbol s and its interpretation *I(s)*. For example, if *f* is a binary function symbol of $\mathcal{A}$, one simply writes $f: \mathcal{A}^2 \to \mathcal{A}$ rather than $f^{\mathcal{A}}: |\mathcal{A}|^2 \to |\mathcal{A}|$.

Examples

The standard signature σ_f for fields consists of two binary function symbols + and ×, where additional symbols can be derived, such as a unary function symbol – (uniquely determined by +) and the two constant symbols 0 and 1 (uniquely determined by + and × respectively). Thus a structure (algebra) for this signature consists of a set of elements *A* together with two binary functions, that can be enhanced with a unary function, and two distinguished elements; but there is no requirement that it satisfy any of the field axioms. The rational numbers *Q*, the real numbers *R* and the complex numbers *C*, like any other field, can be regarded as σ-structures in an obvious way:

$$\mathcal{Q} = (Q, \sigma_f, I_{\mathcal{Q}})$$

$$\mathcal{R} = (R, \sigma_f, I_{\mathcal{R}})$$

$$\mathcal{C} = (C, \sigma_f, I_{\mathcal{C}})$$

In all three cases we have the standard signature given by

$$\sigma_f = (S_f, \mathrm{ar}_f)$$

with

$$S_f = \{+, \times, -, 0, 1\},\ ar_f(+) = ar_f(\times) = 2, ar_f(\ ) = 1, ar_f(0) = ar_f(1) = 0$$

Interpretation functions:

$I_{\mathcal{Q}}(+) : Q \times Q \to Q$ is addition of rational numbers,

$I_{\mathcal{Q}}(\times) : Q \times Q \to Q$ is multiplication of rational numbers,

$I_{\mathcal{Q}}(-) : Q \to Q$ is the function that takes each rational number *x* to -*x*, and

$I_{\mathcal{Q}}(0) \in Q$ is the number 0 and

$I_{\mathcal{Q}}(1) \in Q$ is the number 1;

and $I_{\mathcal{R}}$ and $I_{\mathcal{C}}$ are similarly defined.

But the ring *Z* of integers, which is not a field, is also a σ_f-structure in the same way. In fact, there is no requirement that *any* of the field axioms hold in a σ_f-structure.

A signature for ordered fields needs an additional binary relation such as < or ≤, and therefore structures for such a signature are not algebras, even though they are of course algebraic structures in the usual, loose sense of the word.

The ordinary signature for set theory includes a single binary relation ∈. A structure for this signature consists of a set of elements and an interpretation of the ∈ relation as a binary relation on these elements.

Induced Substructures and Closed Subsets

$\mathcal{A}$ is called an (induced) substructure of $\mathcal{B}$ if

- $\mathcal{A}$ and $\mathcal{B}$ have the same signature $\sigma(\mathcal{A}) = \sigma(\mathcal{B})$;
- the domain of $\mathcal{A}$ is contained in the domain of $\mathcal{B}$: $|\mathcal{A}| \subseteq |\mathcal{B}|$;; and
- the interpretations of all function and relation symbols agree on $|\mathcal{B}|$.

The usual notation for this relation is $\mathcal{A} \subseteq \mathcal{B}$.

A subset $B \subseteq |\mathcal{A}|$ of the domain of a structure $\mathcal{A}$ is called closed if it is closed under the functions of $\mathcal{A}$, i.e. if the following condition is satisfied: for every natural number n, every n-ary function symbol f (in the signature of $\mathcal{A}$) and all elements $b_1, b_2, \ldots, b_n \in B$, the result of applying f to the n-tuple $b_1 b_2 \ldots b_n$ is again an element of B: $f(b_1, b_2, \ldots, b_n) \in B$.

For every subset $B \subseteq |\mathcal{A}|$ there is a smallest closed subset of $|\mathcal{A}|$ that contains B. It is called the closed subset generated by B, or the hull of B, and denoted by $\langle B \rangle$ or $\langle B \rangle_A$. The operator $\langle\rangle$ is a finitary closure operator on the set of subsets of $|\mathcal{A}|$.

If $\mathcal{A} = (A, \sigma, I)$ and $B \subseteq A$ is a closed subset, then (B, σ, I') is an induced substructure of $\mathcal{A}$, where I' assigns to every symbol of σ the restriction to B of its interpretation in $\mathcal{A}$. Conversely, the domain of an induced substructure is a closed subset.

The closed subsets (or induced substructures) of a structure form a lattice. The meet of two subsets is their intersection. The join of two subsets is the closed subset generated by their union. Universal algebra studies the lattice of substructures of a structure in detail.

Examples

Let σ = {+, ×, −, 0, 1} be again the standard signature for fields. When regarded as σ-structures in the natural way, the rational numbers form a substructure of the real numbers, and the real numbers form a substructure of the complex numbers. The rational numbers are the smallest substructure of the real (or complex) numbers that also satisfies the field axioms.

The set of integers gives an even smaller substructure of the real numbers which is not a field. Indeed, the integers are the substructure of the real numbers generated by the empty set, using this signature. The notion in abstract algebra that corresponds to a substructure of a field, in this signature, is that of a subring, rather than that of a subfield.

The most obvious way to define a graph is a structure with a signature σ consisting of a single binary relation symbol E. The vertices of the graph form the domain of the structure, and for two vertices a and b, $(a,b) \in \mathrm{E}$ means that a and b are connected by an edge. In this encoding, the notion of induced substructure is more restrictive than the notion of subgraph. For example, let G be a graph consisting of two vertices connected by an edge, and let H be the graph consisting of the same vertices but no edges. H is a subgraph of G, but not an induced substructure. The notion in graph theory that corresponds to induced substructures is that of induced subgraphs.

Homomorphisms and Embeddings

Homomorphisms

Given two structures $\mathcal{A}$ and $\mathcal{B}$ of the same signature σ, a (σ-)homomorphism from $\mathcal{A}$ to $\mathcal{B}$ is a map $h:|\mathcal{A}|\to|\mathcal{B}|$ that preserves the functions and relations. More precisely:

- For every n-ary function symbol f of σ and any elements $a_1,a_2,\ldots,a_n \in|\mathcal{A}|$, the following equation holds:

  $$h(f(a_1,a_2,\ldots,a_n)) = f(h(a_1),h(a_2),\ldots,h(a_n)).$$

- For every n-ary relation symbol R of σ and any elements $a_1,a_2,\ldots,a_n \in|\mathcal{A}|$, the following implication holds:

  $$(a_1,a_2,\ldots,a_n)\in R \Rightarrow (h(a_1),h(a_2),\ldots,h(a_n))\in R.$$

The notation for a homomorphism h from $\mathcal{A}$ to $\mathcal{B}$ is $h:\mathcal{A}\to\mathcal{B}$.

For every signature σ there is a concrete category σ-Hom which has σ-structures as objects and σ-homomorphisms as morphisms.

A homomorphism $h:\mathcal{A}\to\mathcal{B}$ is sometimes called strong if for every n-ary relation symbol R and any elements $b_1,b_2,\ldots,b_n \in|\mathcal{B}|$ such that $(b_1,b_2,\ldots,b_n)\in R$, there are $a_1,a_2,\ldots,a_n \in|\mathcal{A}|$ such that $(a_1,a_2,\ldots,a_n)\in R$ and $b_1 = h(a_1), b_2 - h(a_2),\ldots, b_n = h(a_n)$. The strong homomorphisms give rise to a subcategory of σ-Hom.

Embeddings

A (σ-)homomorphism $h:\mathcal{A}\to\mathcal{B}$ is called a (σ-)embedding if it is one-to-one and

- for every n-ary relation symbol R of σ and any elements $a_1,a_2,\ldots,a_n$, the following equivalence holds:

  $$(a_1,a_2,\ldots,a_n)\in R \Leftrightarrow (h(a_1),h(a_2),\ldots,h(a_n))\in R.$$

Thus an embedding is the same thing as a strong homomorphism which is one-to-one. The category σ-Emb of σ-structures and σ-embeddings is a concrete subcategory of σ-Hom.

Induced substructures correspond to subobjects in σ-Emb. If σ has only function symbols, σ-Emb is the subcategory of monomorphisms of σ-Hom. In this case induced substructures also correspond to subobjects in σ-Hom.

Example

As seen above, in the standard encoding of graphs as structures the induced substructures are precisely the induced subgraphs. However, a homomorphism between graphs is the same thing as a homomorphism between the two structures coding the graph. In the example of the previous section, even though the subgraph H of G is not induced, the identity map id: $H\to G$ is a homomorphism. This map is in fact a monomorphism in the category σ-Hom, and therefore H is a subobject of G which is not an induced substructure.

Homomorphism Problem

The following problem is known as the *homomorphism problem*:

> Given two finite structures $\mathcal{A}$ and $\mathcal{B}$ of a finite relational signature, find a homomorphism $h:\mathcal{A}\rightarrow\mathcal{B}$ or show that no such homomorphism exists.

Every constraint satisfaction problem (CSP) has a translation into the homomorphism problem. Therefore, the complexity of CSP can be studied using the methods of finite model theory.

Another application is in database theory, where a relational model of a database is essentially the same thing as a relational structure. It turns out that a conjunctive query on a database can be described by another structure in the same signature as the database model. A homomorphism from the relational model to the structure representing the query is the same thing as a solution to the query. This shows that the conjunctive query problem is also equivalent to the homomorphism problem.

Structures and First-order Logic

Structures are sometimes referred to as "first-order structures". This is misleading, as nothing in their definition ties them to any specific logic, and in fact they are suitable as semantic objects both for very restricted fragments of first-order logic such as that used in universal algebra, and for second-order logic. In connection with first-order logic and model theory, structures are often called models, even when the question "models of what?" has no obvious answer.

Satisfaction Relation

Each first-order structure $\mathcal{M}$ has a satisfaction relation $\mathcal{M}\vDash\phi$ defined for all formulas ϕ in the language consisting of the language of $\mathcal{M}$ together with a constant symbol for each element of *M*, which is interpreted as that element. This relation is defined inductively using Tarski's T-schema.

A structure $\mathcal{M}$ is said to be a model of a theory *T* if the language of $\mathcal{M}$ is the same as the language of *T* and every sentence in *T* is satisfied by $\mathcal{M}$. Thus, for example, a "ring" is a structure for the language of rings that satisfies each of the ring axioms, and a model of ZFC set theory is a structure in the language of set theory that satisfies each of the ZFC axioms.

Definable Relations

An *n*-ary relation *R* on the universe *M* of a structure $\mathcal{M}$ is said to be definable (or explicitly definable, or $\varnothing$-definable) if there is a formula $\varphi(x_1,...,x_n)$ such that

$$R=\{(a_1,\ldots,a_n)\in M^n:\mathcal{M}\vDash\varphi(a_1,\ldots,a_n)\}.$$

In other words, *R* is definable if and only if there is a formula φ such that

$$(a_1,\ldots,a_n)\in R\Leftrightarrow\mathcal{M}\vDash\varphi(a_1,\ldots,a_n)$$

is correct.

An important special case is the definability of specific elements. An element m of M is definable in $\mathcal{M}$ if and only if there is a formula φ(x) such that

$$\mathcal{M} \vDash \forall x(x = m \leftrightarrow \varphi(x)).$$

Definability with Parameters

A relation R is said to be definable with parameters (or $|\mathcal{M}|$-definable) if there is a formula φ with parameters from $\mathcal{M}$ such that R is definable using φ. Every element of a structure is definable using the element itself as a parameter.

It should be noted that some authors use *definable* to mean *definable without parameters*, while other authors mean *definable with parameters*. Broadly speaking, the convention that *definable* means *definable without parameters* is more common amongst set theorists, while the opposite convention is more common amongst model theorists.

Implicit Definability

Recall from above that an n-ary relation R on the universe M of a structure is explicitly definable if there is a formula $\varphi(x_1,...,x_n)$ such that

$$R = \{(a_1,\ldots,a_n) \in M^n : \mathcal{M} \vDash \varphi(a_1,\ldots,a_n)\}.$$

Here the formula φ used to define a relation R must be over the signature of $\mathcal{M}$ and so φ may not mention R itself, since R is not in the signature of $\mathcal{M}$. If there is a formula φ in the extended language containing the language of $\mathcal{M}$ and a new symbol R, and the relation R is the only relation on $\mathcal{M}$ such that $\mathcal{M} \vDash \varphi$, then R is said to be implicitly definable over $\mathcal{M}$.

By Beth's theorem, every implicitly definable relation is explicitly definable.

Many-sorted Structures

Structures as defined above are sometimes called one-sorted structures to distinguish them from the more general many-sorted structures. A many-sorted structure can have an arbitrary number of domains. The sorts are part of the signature, and they play the role of names for the different domains. Many-sorted signatures also prescribe on which sorts the functions and relations of a many-sorted structure are defined. Therefore, the arities of function symbols or relation symbols must be more complicated objects such as tuples of sorts rather than natural numbers.

Vector spaces, for example, can be regarded as two-sorted structures in the following way. The two-sorted signature of vector spaces consists of two sorts V (for vectors) and S (for scalars) and the following function symbols:

- $+_S$ and $\times_S$ of arity (S, S; S).
- $-_S$ of arity (S; S).
- 0_S and 1_S of arity (S).
- $+_V$ of arity (V, V; V).
- $-_V$ of arity (V; V).
- 0_V of arity (V).
- $\times$ of arity (S, V; V).

If V is a vector space over a field F, the corresponding two-sorted structure $\mathcal{V}$ consists of the vector domain $|\mathcal{V}|_V = V$, the scalar domain $|\mathcal{V}|_S = F$, and the obvious functions, such as the vector zero $0_V^{\mathcal{V}} = 0 \in |\mathcal{V}|_V$, the scalar zero $0_S^{\mathcal{V}} = 0 \in |\mathcal{V}|_S$, or scalar multiplication $\times^{\mathcal{V}} : |\mathcal{V}|_S \times |\mathcal{V}|_V \to |\mathcal{V}|_V$.

Many-sorted structures are often used as a convenient tool even when they could be avoided with a little effort. But they are rarely defined in a rigorous way, because it is straightforward and tedious (hence unrewarding) to carry out the generalization explicitly.

In most mathematical endeavours, not much attention is paid to the sorts. A many-sorted logic however naturally leads to a type theory. As Bart Jacobs puts it: "A logic is always a logic over a type theory." This emphasis in turn leads to categorical logic because a logic over a type theory categorically corresponds to one ("total") category, capturing the logic, being fibred over another ("base") category, capturing the type theory.

Other Generalizations

Partial Algebras

Both universal algebra and model theory study classes of (structures or) algebras that are defined by a signature and a set of axioms. In the case of model theory these axioms have the form of first-order sentences. The formalism of universal algebra is much more restrictive; essentially it only allows first-order sentences that have the form of universally quantified equations between terms, e.g. $\forall x \forall y \, (x + y = y + x)$. One consequence is that the choice of a signature is more significant in universal algebra than it is in model theory. For example, the class of groups, in the signature consisting of the binary function symbol × and the constant symbol 1, is an elementary class, but it is not a variety. Universal algebra solves this problem by adding a unary function $symbol^{-1}$.

In the case of fields this strategy works only for addition. For multiplication it fails because 0 does not have a multiplicative inverse. An ad hoc attempt to deal with this would be to define $0^{-1} = 0$. (This attempt fails, essentially because with this definition $0 \times 0^{-1} = 1$ is not true.) Therefore, one is naturally led to allow partial functions, i.e., functions that are defined only on a subset of their domain. However, there are several obvious ways to generalize notions such as substructure, homomorphism and identity.

Structures for Typed Languages

In type theory, there are many sorts of variables, each of which has a type. Types are inductively defined; given two types δ and σ there is also a type $\sigma \to \delta$ that represents functions from objects of type σ to objects of type δ. A structure for a typed language (in the ordinary first-order semantics) must include a separate set of objects of each type, and for a function type the structure must have complete information about the function represented by each object of that type.

Higher-order Languages

There is more than one possible semantics for higher-order logic, as discussed in the article on second-order logic. When using full higher-order semantics, a structure need only have a universe for objects of type o, and the T-schema is extended so that a quantifier over a higher-order type is satisfied by the model if and only if it is disquotationally true. When using first-order semantics, an additional sort is added for each higher-order type, as in the case of a many sorted first order language.

Structures that are Proper Classes

In the study of set theory and category theory, it is sometimes useful to consider structures in which the domain of discourse is a proper class instead of a set. These structures are sometimes called class models to distinguish them from the "set models" discussed above. When the domain is a proper class, each function and relation symbol may also be represented by a proper class.

In Bertrand Russell's *Principia Mathematica*, structures were also allowed to have a proper class as their domain

Subalgebra

In mathematics, a subalgebra is a subset of an algebra, closed under all its operations, and carrying the induced operations.

"Algebra", when referring to a structure, often means a vector space or module equipped with an additional bilinear operation. Algebras in universal algebra are far more general: they are a common generalisation of *all* algebraic structures. Subalgebra can be a subset of both cases.

Subalgebras for Algebras over a Ring or Field

A subalgebra of an algebra over a commutative ring or field is a vector subspace which is closed under the multiplication of vectors. The restriction of the algebra multiplication makes it an algebra over the same ring or field. This notion also applies to most specializations, where the multiplication must satisfy additional properties, e.g. to associative algebras or to Lie algebras. Only for unital algebras is there a stronger notion, of unital subalgebra, for which it is also required that the unit of the subalgebra be the unit of the bigger algebra.

Example

The 2×2-matrices over the reals form a unital algebra in the obvious way. The 2×2-matrices for which all entries are zero, except for the first one on the diagonal, form a subalgebra. It is also unital, but it is not a unital subalgebra.

Subalgebras in Universal Algebra

In universal algebra, a subalgebra of an algebra A is a subset S of A that also has the structure of an algebra of the same type when the algebraic operations are restricted to S. If the axioms of a kind of algebraic structure is described by equational laws, as is typically the case in universal algebra, then the only thing that needs to be checked is that S is *closed* under the operations.

Some authors consider algebras with partial functions. There are various ways of defining subalgebras for these. Another generalization of algebras is to allow relations. These more general algebras are usually called structures, and they are studied in model theory and in theoretical computer science. For structures with relations there are notions of weak and of induced substructures.

Example

For example, the standard signature for groups in universal algebra is $(\bullet, ^{-1}, 1)$. (Inversion and unit are needed to get the right notions of homomorphism and so that the group laws can be expressed as equations.) Therefore, a subgroup of a group G is a subset S of G such that:

- the identity e of G belongs to S (so that S is closed under the identity constant operation);
- whenever x belongs to S, so does x^{-1} (so that S is closed under the inverse operation);
- whenever x and y belong to S, so does $x \bullet y$ (so that S is closed under the group's multiplication operation).

Substructure

In mathematical logic, an (induced) substructure or (induced) subalgebra is a structure whose domain is a subset of that of a bigger structure, and whose functions and relations are the traces of the functions and relations of the bigger structure. Some examples of subalgebras are subgroups, submonoids, subrings, subfields, subalgebras of algebras over a field, or induced subgraphs. Shifting the point of view, the larger structure is called an extension or a superstructure of its substructure.

In model theory, the term "submodel" is often used as a synonym for substructure, especially when the context suggests a theory of which both structures are models.

In the presence of relations (i.e. for structures such as ordered groups or graphs, whose signature is not functional) it may make sense to relax the conditions on a subalgebra so that the relations on a weak substructure (or weak subalgebra) are *at most* those induced from the bigger structure. Subgraphs are an example where the distinction matters, and the term "subgraph" does indeed refer to weak substructures. Ordered groups, on the other hand, have the special property that every substructure of an ordered group which is itself an ordered group, is an induced substructure.

Definition

Given two structures A and B of the same signature σ, A is said to be a weak substructure of B, or a weak subalgebra of B, if

- the domain of A is a subset of the domain of B,
- $f^A = f^B|_{A^n}$ for every n-ary function symbol f in σ, and
- $R^A \subseteq R^B \cap A^n$ for every n-ary relation symbol R in σ.

A is said to be a substructure of B, or a subalgebra of B, if A is a weak subalgebra of B and, moreover,

- $R^A = R^B \cap A^n$ for every n-ary relation symbol R in σ.

If A is a substructure of B, then B is called a superstructure of A or, especially if A is an induced substructure, an extension of A.

Example

In the language consisting of the binary functions + and ×, binary relation <, and constants 0 and 1, the structure (Q, +, ×, <, 0, 1) is a substructure of (R, +, ×, <, 0, 1). More generally, the substructures of an ordered field (or just a field) are precisely its subfields. Similarly, in the language $\left(\times, ^{-1}, 1\right)$ of groups, the substructures of a group are its subgroups. In the language (×, 1) of monoids, however, the substructures of a group are its submonoids. They need not be groups; and even if they are groups, they need not be subgroups.

In the case of graphs (in the signature consisting of one binary relation), subgraphs, and its weak substructures are precisely its subgraphs.

As Subobjects

For every signature σ, induced substructures of σ-structures are the subobjects in the concrete category of σ-structures and strong homomorphisms (and also in the concrete category of σ-structures and σ-embeddings). Weak substructures of σ-structures are the subobjects in the concrete category of σ-structures and homomorphisms in the ordinary sense.

Submodel

In model theory, given a structure M which is a model of a theory T, a submodel of M in a narrower sense is a substructure of M which is also a model of T. For example, if T is the theory of abelian groups in the signature (+, 0), then the submodels of the group of integers (Z, +, 0) are the substructures which are also abelian groups. Thus the natural numbers (N, +, 0) form a substructure of (Z, +, 0) which is not a submodel, while the even numbers (2Z, +, 0) form a submodel.

Other examples:

1. The algebraic numbers form a submodel of the complex numbers in the theory of algebraically closed fields.
2. The rational numbers form a submodel of the real numbers in the theory of fields.
3. Every elementary substructure of a model of a theory T also satisfies T; hence it is a submodel.

In the category of models of a theory and embeddings between them, the submodels of a model are its subobjects.

Subdirectly Irreducible Algebra

In the branch of mathematics known as universal algebra (and in its applications), a subdirectly irreducible algebra is an algebra that cannot be factored as a subdirect product of "simpler" algebras. Subdirectly irreducible algebras play a somewhat analogous role in algebra to primes in number theory.

Definition

A universal algebra A is said to be subdirectly irreducible when A has more than one element, and when any subdirect representation of A includes (as a factor) an algebra isomorphic to A, with the isomorphism being given by the projection map.

Examples

- The two-element chain, as either a Boolean algebra, a Heyting algebra, a lattice, or a semilattice, is subdirectly irreducible. In fact, a distributive lattice is subdirectly irreducible if and only if it has exactly two elements.
- Any finite chain with two or more elements, as a Heyting algebra, is subdirectly irreducible. (This is not the case for chains of three or more elements as either lattices or semilattices, which are subdirectly reducible to the two-element chain. The difference with Heyting algebras is that $a \to b$ need not be comparable with a under the lattice order even when b is.)
- Any finite cyclic group of order a power of a prime (i.e. any finite p-group) is subdirectly irreducible. (One weakness of the analogy between subdirect irreducibles and prime numbers is that the integers are subdirectly representable by any infinite family of nonisomorphic prime-power cyclic groups, e.g. just those of order a Mersenne prime assuming there are infinitely many.) In fact, an abelian group is subdirectly irreducible if and only if it is isomorphic to a finite p-group or isomorphic to a Prüfer group (an infinite but countable p-group, which is the direct limit of its finite p-subgroups).
- A vector space is subdirectly irreducible if and only if it has dimension one.

Properties

The subdirect representation theorem of universal algebra states that every algebra is subdirectly representable by its subdirectly irreducible quotients. An equivalent definition of "subdirect irreducible" therefore is any algebra A that is not subdirectly representable by those of its quotients not isomorphic to A. (This is not quite the same thing as "by its proper quotients" because a proper quotient of A may be isomorphic to A, for example the quotient of the semilattice (Z, min) obtained by identifying just the two elements 3 and 4.)

An immediate corollary is that any variety, as a class closed under homomorphisms, subalgebras, and direct products, is determined by its subdirectly irreducible members, since every algebra A in the variety can be constructed as a subalgebra of a suitable direct product of the subdirectly irreducible quotients of *A*, all of which belong to the variety because *A* does. For this reason one often studies not the variety itself but just its subdirect irreducibles.

An algebra *A* is subdirectly irreducible if and only if it contains two elements that are identified by every proper quotient, equivalently, if and only if its lattice Con *A* of congruences has a least nonidentity element. That is, any subdirect irreducible must contain a specific pair of elements witnessing its irreducibility in this way. Given such a witness (a,b) to subdirect irreducibility we say that the subdirect irreducible is (a,b)-irreducible.

Given any class *C* of similar algebras, Jónsson's Lemma (due to Bjarni Jónsson) states that if the variety HSP(*C*) generated by *C* is congruence-distributive, its subdirect irreducibles are in $HSP_U(C)$, that is, they are quotients of subalgebras of ultraproducts of members of *C*. (If *C* is a finite set of finite algebras, the ultraproduct operation is redundant.)

Applications

A necessary and sufficient condition for a Heyting algebra to be subdirectly irreducible is for there to be a greatest element strictly below 1. The witnessing pair is that element and 1, and identifying any other pair *a*, *b* of elements identifies both $a \to b$ and $b \to a$ with 1 thereby collapsing everything above those two implications to 1. Hence every finite chain of two or more elements as a Heyting algebra is subdirectly irreducible.

By Jónsson's Lemma, subdirectly irreducible algebras of a congruence-distributive variety generated by a finite set of finite algebras are no larger than the generating algebras, since the quotients and subalgebras of an algebra *A* are never larger than *A* itself. For example, the subdirect irreducibles in the variety generated by a finite linearly ordered Heyting algebra *H* must be just the nondegenerate quotients of *H*, namely all smaller linearly ordered nondegenerate Heyting algebras. The conditions cannot be dropped in general: for example, the variety of all Heyting algebras is generated by the set of its finite subdirectly irreducible algebras, but there exist subdirectly irreducible Heyting algebras of arbitrary (infinite) cardinality. There also exists a single finite algebra generating a (non-congruence-distributive) variety with arbitrarily large subdirect irreducibles.

References

- Arai, Yoshinari; Iséki, Kiyoshi; Tanaka, Shôtarô (1966), "Characterizations of BCI, BCK-algebras", Proc. Japan Acad., 42 (2): 105–107, doi:10.3792/pja/1195522126, MR 0202572
- Bergman, George M., 1998. An Invitation to General Algebra and Universal Constructions (pub. Henry Helson, 15 the Crescent, Berkeley CA, 94708) 398 pp. ISBN 0-9655211-4-1
- Angell, R. B. (1970), "Review of several papers on BCI, BCK-Algebras", The Journal of Symbolic Logic, 35 (3): 465–466, doi:10.2307/2270728, ISSN 0022-4812, JSTOR 2270728
- Burris, Stanley N., and H.P. Sankappanavar, 1981. A Course in Universal Algebra Springer-Verlag. ISBN 3-540-90578-2
- Imai, Y.; Iséki, K (1966), "On axiom systems of propositional calculi, XIV", Proc. Japan Acad. Ser. A, Math. Sci., 42: 19–22, doi:10.3792/pja/1195522169

4

Integrating Algebra and Geometry

Algebraic geometry is a branch of mathematics that studies zeros of multivariate polynomials. Some of the classes of algebraic varieties that are commonly studied are plane algebraic curves, cubic curves and quartic curves. The topics elaborated in this chapter will help in gaining a better perspective about the different concepts that are central to algebraic geometry.

Algebraic Geometry

Algebraic geometry is a branch of mathematics, classically studying zeros of multivariate polynomials. Modern algebraic geometry is based on the use of abstract algebraic techniques, mainly from commutative algebra, for solving geometrical problems about these sets of zeros.

The fundamental objects of study in algebraic geometry are algebraic varieties, which are geometric manifestations of solutions of systems of polynomial equations. Examples of the most studied classes of algebraic varieties are: plane algebraic curves, which include lines, circles, parabolas, ellipses, hyperbolas, cubic curves like elliptic curves, and quartic curves like lemniscates and Cassini ovals. A point of the plane belongs to an algebraic curve if its coordinates satisfy a given polynomial equation. Basic questions involve the study of the points of special interest like the singular points, the inflection points and the points at infinity. More advanced questions involve the topology of the curve and relations between the curves given by different equations.

Algebraic geometry occupies a central place in modern mathematics and has multiple conceptual connections with such diverse fields as complex analysis, topology and number theory. Initially a study of systems of polynomial equations in several variables, the subject of algebraic geometry starts where equation solving leaves off, and it becomes even more important to understand the intrinsic properties of the totality of solutions of a system of equations, than to find a specific solution; this leads into some of the deepest areas in all of mathematics, both conceptually and in terms of technique.

In the 20th century, algebraic geometry split into several subareas.

- The mainstream of algebraic geometry is devoted to the study of the complex points of the algebraic varieties and more generally to the points with coordinates in an algebraically closed field.
- The study of the points of an algebraic variety with coordinates in the field of the rational numbers or in a number field became arithmetic geometry (or more classically Diophantine geometry), a subfield of algebraic number theory.
- The study of the real points of an algebraic variety is the subject of real algebraic geometry.

- A large part of singularity theory is devoted to the singularities of algebraic varieties.
- With the rise of the computers, a computational algebraic geometry area has emerged, which lies at the intersection of algebraic geometry and computer algebra. It consists essentially in developing algorithms and software for studying and finding the properties of explicitly given algebraic varieties.

Much of the development of the mainstream of algebraic geometry in the 20th century occurred within an abstract algebraic framework, with increasing emphasis being placed on "intrinsic" properties of algebraic varieties not dependent on any particular way of embedding the variety in an ambient coordinate space; this parallels developments in topology, differential and complex geometry. One key achievement of this abstract algebraic geometry is Grothendieck's scheme theory which allows one to use sheaf theory to study algebraic varieties in a way which is very similar to its use in the study of differential and analytic manifolds. This is obtained by extending the notion of point: In classical algebraic geometry, a point of an affine variety may be identified, through Hilbert's Nullstellensatz, with a maximal ideal of the coordinate ring, while the points of the corresponding affine scheme are all prime ideals of this ring. This means that a point of such a scheme may be either a usual point or a subvariety. This approach also enables a unification of the language and the tools of classical algebraic geometry, mainly concerned with complex points, and of algebraic number theory. Wiles's proof of the longstanding conjecture called Fermat's last theorem is an example of the power of this approach.

Basic Notions

Zeros of Simultaneous Polynomials

In classical algebraic geometry, the main objects of interest are the vanishing sets of collections of polynomials, meaning the set of all points that simultaneously satisfy one or more polynomial equations. For instance, the two-dimensional sphere of radius 1 in three-dimensional Euclidean space R^3 could be defined as the set of all points (x,y,z) with

$$x^2 + y^2 + z^2 - 1 = 0.$$

A "slanted" circle in R^3 can be defined as the set of all points (x,y,z) which satisfy the two polynomial equations

$$x^2 + y^2 + z^2 - 1 = 0,$$
$$x + y + z = 0.$$

Affine Varieties

First we start with a field k. In classical algebraic geometry, this field was always the complex numbers C, but many of the same results are true if we assume only that k is algebraically closed. We consider the affine space of dimension n over k, denoted $A^n(k)$ (or more simply A^n, when k is clear from the context). When one fixes a coordinate system, one may identify $A^n(k)$ with k^n. The purpose of not working with k^n is to emphasize that one "forgets" the vector space structure that k^n carries.

A function $f : \mathrm{A}^n \rightarrow \mathrm{A}^1$ is said to be *polynomial* (or *regular*) if it can be written as a polynomial, that is, if there is a polynomial p in $k[x_1,...,x_n]$ such that $f(M) = p(t_1,...,t_n)$ for every point M with coordinates $(t_1,...,t_n)$ in A^n. The property of a function to be polynomial (or regular) does not depend on the choice of a coordinate system in A^n.

When a coordinate system is chosen, the regular functions on the affine n-space may be identified with the ring of polynomial functions in n variables over k. Therefore, the set of the regular functions on A^n is a ring, which is denoted $k[\mathrm{A}^n]$.

We say that a polynomial *vanishes* at a point if evaluating it at that point gives zero. Let S be a set of polynomials in $k[\mathrm{A}^n]$. The *vanishing set of S* (or *vanishing locus* or *zero set*) is the set $V(S)$ of all points in A^n where every polynomial in S vanishes. Symbolically,

$$V(S) = \{(t_1,\ldots,t_n) \mid \forall p \in S, p(t_1,\ldots,t_n) = 0\}.$$

A subset of A^n which is $V(S)$, for some S, is called an *algebraic set*. The V stands for *variety* (a specific type of algebraic set to be defined below).

Given a subset U of A^n, can one recover the set of polynomials which generate it? If U is *any* subset of A^n, define $I(U)$ to be the set of all polynomials whose vanishing set contains U. The I stands for ideal: if two polynomials f and g both vanish on U, then $f+g$ vanishes on U, and if h is any polynomial, then hf vanishes on U, so $I(U)$ is always an ideal of the polynomial ring $k[\mathrm{A}^n]$.

Two natural questions to ask are:

- Given a subset U of A^n, when is $U = V(I(U))$?
- Given a set S of polynomials, when is $S = I(V(S))$?

The answer to the first question is provided by introducing the Zariski topology, a topology on A^n whose closed sets are the algebraic sets, and which directly reflects the algebraic structure of $k[\mathrm{A}^n]$. Then $U = V(I(U))$ if and only if U is an algebraic set or equivalently a Zariski-closed set. The answer to the second question is given by Hilbert's Nullstellensatz. In one of its forms, it says that $I(V(S))$ is the radical of the ideal generated by S. In more abstract language, there is a Galois connection, giving rise to two closure operators; they can be identified, and naturally play a basic role in the theory; the example is elaborated at Galois connection.

For various reasons we may not always want to work with the entire ideal corresponding to an algebraic set U. Hilbert's basis theorem implies that ideals in $k[\mathrm{A}^n]$ are always finitely generated.

An algebraic set is called *irreducible* if it cannot be written as the union of two smaller algebraic sets. Any algebraic set is a finite union of irreducible algebraic sets and this decomposition is unique. Thus its elements are called the *irreducible components* of the algebraic set. An irreducible algebraic set is also called a *variety*. It turns out that an algebraic set is a variety if and only if it may be defined as the vanishing set of a prime ideal of the polynomial ring.

Some authors do not make a clear distinction between algebraic sets and varieties and use *irreducible variety* to make the distinction when needed.

Regular Functions

Just as continuous functions are the natural maps on topological spaces and smooth functions are the natural maps on differentiable manifolds, there is a natural class of functions on an algebraic set, called *regular functions* or *polynomial functions*. A regular function on an algebraic set V contained in A^n is the restriction to V of a regular function on A^n. For an algebraic set defined on the field of the complex numbers, the regular functions are smooth and even analytic.

It may seem unnaturally restrictive to require that a regular function always extend to the ambient space, but it is very similar to the situation in a normal topological space, where the Tietze extension theorem guarantees that a continuous function on a closed subset always extends to the ambient topological space.

Just as with the regular functions on affine space, the regular functions on V form a ring, which we denote by $k[V]$. This ring is called the *coordinate ring of V*.

Since regular functions on V come from regular functions on A^n, there is a relationship between the coordinate rings. Specifically, if a regular function on V is the restriction of two functions f and g in $k[A^n]$, then $f - g$ is a polynomial function which is null on V and thus belongs to $I(V)$. Thus $k[V]$ may be identified with $k[A^n]/I(V)$.

Morphism of Affine Varieties

Using regular functions from an affine variety to A^1, we can define regular maps from one affine variety to another. First we will define a regular map from a variety into affine space: Let V be a variety contained in A^n. Choose m regular functions on V, and call them $f_1, ..., f_m$. We define a *regular map* f from V to A^m by letting $f = (f_1, ..., f_m)$. In other words, each f_i determines one coordinate of the range of f.

If V' is a variety contained in A^m, we say that f is a *regular map* from V to V' if the range of f is contained in V'.

The definition of the regular maps apply also to algebraic sets. The regular maps are also called *morphisms*, as they make the collection of all affine algebraic sets into a category, where the objects are the affine algebraic sets and the morphisms are the regular maps. The affine varieties is a subcategory of the category of the algebraic sets.

Given a regular map g from V to V' and a regular function f of $k[V']$, then $f \circ g \in k[V]$. The map $f \rightarrow f \circ g$ is a ring homomorphism from $k[V']$ to $k[V]$. Conversely, every ring homomorphism from $k[V']$ to $k[V]$ defines a regular map from V to V'. This defines an equivalence of categories between the category of algebraic sets and the opposite category of the finitely generated reduced k-algebras. This equivalence is one of the starting points of scheme theory.

Rational Function and Birational Equivalence

In contrast to the preceding sections, this section concerns only varieties and not algebraic sets. On the other hand, the definitions extend naturally to projective varieties (next section), as an affine variety and its projective completion have the same field of functions.

If V is an affine variety, its coordinate ring is an integral domain and has thus a field of fractions which is denoted $k(V)$ and called the *field of the rational functions* on V or, shortly, the *function field* of V. Its elements are the restrictions to V of the rational functions over the affine space containing V. The domain of a rational function f is not V but the complement of the subvariety (a hypersurface) where the denominator of f vanishes.

As with regular maps, one may define a *rational map* from a variety V to a variety V'. As with the regular maps, the rational maps from V to V' may be identified to the field homomorphisms from $k(V')$ to $k(V)$.

Two affine varieties are *birationally equivalent* if there are two rational functions between them which are inverse one to the other in the regions where both are defined. Equivalently, they are birationally equivalent if their function fields are isomorphic.

An affine variety is a *rational variety* if it is birationally equivalent to an affine space. This means that the variety admits a rational parameterization. For example, the circle of equation $x^2 + y^2 - 1 = 0$ is a rational curve, as it has the parameterization

$$x = \frac{2t}{1+t^2}$$

$$y = \frac{1-t^2}{1+t^2},$$

which may also be viewed as a rational map from the line to the circle.

The problem of resolution of singularities is to know if every algebraic variety is birationally equivalent to a variety whose projective completion is nonsingular. It was solved in the affirmative in characteristic 0 by Heisuke Hironaka in 1964 and is yet unsolved in finite characteristic.

Projective Variety

Just as the formulas for the roots of second, third, and fourth degree polynomials suggest extending real numbers to the more algebraically complete setting of the complex numbers, many properties of algebraic varieties suggest extending affine space to a more geometrically complete projective space. Whereas the complex numbers are obtained by adding the number i, a root of the polynomial $x^2 + 1$, projective space is obtained by adding in appropriate points "at infinity", points where parallel lines may meet.

To see how this might come about, consider the variety $V(y - x^2)$. If we draw it, we get a parabola. As x goes to positive infinity, the slope of the line from the origin to the point (x, x^2) also goes to positive infinity. As x goes to negative infinity, the slope of the same line goes to negative infinity.

Compare this to the variety $V(y - x^3)$. This is a cubic curve. As x goes to positive infinity, the slope of the line from the origin to the point (x, x^3) goes to positive infinity just as before. But unlike before, as x goes to negative infinity, the slope of the same line goes to positive infinity as well; the exact opposite of the parabola. So the behavior "at infinity" of $V(y - x^3)$ is different from the behavior "at infinity" of $V(y - x^2)$.

The consideration of the *projective completion* of the two curves, which is their prolongation "at infinity" in the projective plane, allows us to quantify this difference: the point at infinity of the parabola is a regular point, whose tangent is the line at infinity, while the point at infinity of the cubic curve is a cusp. Also, both curves are rational, as they are parameterized by x, and the Riemann-Roch theorem implies that the cubic curve must have a singularity, which must be at infinity, as all its points in the affine space are regular.

Thus many of the properties of algebraic varieties, including birational equivalence and all the topological properties, depend on the behavior "at infinity" and so it is natural to study the varieties in projective space. Furthermore, the introduction of projective techniques made many theorems in algebraic geometry simpler and sharper: For example, Bézout's theorem on the number of intersection points between two varieties can be stated in its sharpest form only in projective space. For these reasons, projective space plays a fundamental role in algebraic geometry.

Nowadays, the *projective space* P^n of dimension n is usually defined as the set of the lines passing through a point, considered as the origin, in the affine space of dimension $n + 1$, or equivalently to the set of the vector lines in a vector space of dimension $n + 1$. When a coordinate system has been chosen in the space of dimension $n + 1$, all the points of a line have the same set of coordinates, up to the multiplication by an element of k. This defines the homogeneous coordinates of a point of P^n as a sequence of $n + 1$ elements of the base field k, defined up to the multiplication by a nonzero element of k (the same for the whole sequence).

A polynomial in $n + 1$ variables vanishes at all points of a line passing through the origin if and only if it is homogeneous. In this case, one says that the polynomial *vanishes* at the corresponding point of P^n. This allows us to define a *projective algebraic set* in P^n as the set $V(f_1, ..., f_k)$, where a finite set of homogeneous polynomials $\{f_1, ..., f_k\}$ vanishes. Like for affine algebraic sets, there is a bijection between the projective algebraic sets and the reduced homogeneous ideals which define them. The *projective varieties* are the projective algebraic sets whose defining ideal is prime. In other words, a projective variety is a projective algebraic set, whose homogeneous coordinate ring is an integral domain, the *projective coordinates ring* being defined as the quotient of the graded ring or the polynomials in $n + 1$ variables by the homogeneous (reduced) ideal defining the variety. Every projective algebraic set may be uniquely decomposed into a finite union of projective varieties.

The only regular functions which may be defined properly on a projective variety are the constant functions. Thus this notion is not used in projective situations. On the other hand, the *field of the rational functions* or *function field* is a useful notion, which, similarly to the affine case, is defined as the set of the quotients of two homogeneous elements of the same degree in the homogeneous coordinate ring.

Real Algebraic Geometry

Real algebraic geometry is the study of the real points of algebraic geometry.

The fact that the field of the real numbers is an ordered field cannot be ignored in such a study. For example, the curve of equation $x^2 + y^2 - a = 0$ is a circle if $a > 0$, but does not have any real point if $a < 0$. It follows that real algebraic geometry is not only the study of the real algebraic varieties, but has been generalized to the study of the *semi-algebraic sets*, which are the solutions of systems of polynomial equations and polynomial inequalities. For example, a branch of the hyperbola of equation $xy-1=0$ is not an algebraic variety, but is a semi-algebraic set defined by $xy-1=0$ and $x>0$ or by $xy-1=0$ and $x+y>0$.

One of the challenging problems of real algebraic geometry is the unsolved Hilbert's sixteenth problem: Decide which respective positions are possible for the ovals of a nonsingular plane curve of degree 8.

Computational Algebraic Geometry

One may date the origin of computational algebraic geometry to meeting EUROSAM'79 (International Symposium on Symbolic and Algebraic Manipulation) held at Marseille, France in June 1979. At this meeting,

- Dennis S. Arnon showed that George E. Collins's Cylindrical algebraic decomposition (CAD) allows the computation of the topology of semi-algebraic sets,
- Bruno Buchberger presented the Gröbner bases and his algorithm to compute them,
- Daniel Lazard presented a new algorithm for solving systems of homogeneous polynomial equations with a computational complexity which is essentially polynomial in the expected number of solutions and thus simply exponential in the number of the unknowns. This algorithm is strongly related with Macaulay's multivariate resultant.

Since then, most results in this area are related to one or several of these items either by using or improving one of these algorithms, or by finding algorithms whose complexity is simply exponential in the number of the variables.

A body of mathematical theory complementary to symbolic methods called numerical algebraic geometry has been developed over the last several decades. The main computational method is homotopy continuation. This supports, for example, a model of floating point computation for solving problems of algebraic geometry.

Gröbner Basis

A Gröbner basis is a system of generators of a polynomial ideal whose computation allows the deduction of many properties of the affine algebraic variety defined by the ideal.

Given an ideal I defining an algebraic set V:

- V is empty (over an algebraically closed extension of the basis field), if and only if the Gröbner basis for any monomial ordering is reduced to {1}.
- By means of the Hilbert series one may compute the dimension and the degree of V from any Gröbner basis of I for a monomial ordering refining the total degree.
- If the dimension of V is 0, one may compute the points (finite in number) of V from any Gröbner basis of I.
- A Gröbner basis computation allows one to remove from V all irreducible components which are contained in a given hyper-surface.

- A Gröbner basis computation allows one to compute the Zariski closure of the image of V by the projection on the k first coordinates, and the subset of the image where the projection is not proper.
- More generally Gröbner basis computations allow one to compute the Zariski closure of the image and the critical points of a rational function of V into another affine variety.

Gröbner basis computations do not allow one to compute directly the primary decomposition of I nor the prime ideals defining the irreducible components of V, but most algorithms for this involve Gröbner basis computation. The algorithms which are not based on Gröbner bases use regular chains but may need Gröbner bases in some exceptional situations.

Gröbner bases are deemed to be difficult to compute. In fact they may contain, in the worst case, polynomials whose degree is doubly exponential in the number of variables and a number of polynomials which is also doubly exponential. However, this is only a worst case complexity, and the complexity bound of Lazard's algorithm of 1979 may frequently apply. Faugère F5 algorithm realizes this complexity, as it may be viewed as an improvement of Lazard's 1979 algorithm. It follows that the best implementations allow one to compute almost routinely with algebraic sets of degree more than 100. This means that, presently, the difficulty of computing a Gröbner basis is strongly related to the intrinsic difficulty of the problem.

Cylindrical Algebraic Decomposition (CAD)

CAD is an algorithm which was introduced in 1973 by G. Collins to implement with an acceptable complexity the Tarski–Seidenberg theorem on quantifier elimination over the real numbers.

This theorem concerns the formulas of the first-order logic whose atomic formulas are polynomial equalities or inequalities between polynomials with real coefficients. These formulas are thus the formulas which may be constructed from the atomic formulas by the logical operators *and* (∧), *or* (∨), *not* ( ), *for all* (∀) and *exists* (∃). Tarski's theorem asserts that, from such a formula, one may compute an equivalent formula without quantifier (∀, ∃).

The complexity of CAD is doubly exponential in the number of variables. This means that CAD allows, in theory, to solve every problem of real algebraic geometry which may be expressed by such a formula, that is almost every problem concerning explicitly given varieties and semi-algebraic sets.

While Gröbner basis computation has doubly exponential complexity only in rare cases, CAD has almost always this high complexity. This implies that, unless if most polynomials appearing in the input are linear, it may not solve problems with more than four variables.

Since 1973, most of the research on this subject is devoted either to improve CAD or to find alternate algorithms in special cases of general interest.

As an example of the state of art, there are efficient algorithms to find at least a point in every connected component of a semi-algebraic set, and thus to test if a semi-algebraic set is empty. On the other hand, CAD is yet, in practice, the best algorithm to count the number of connected components.

Asymptotic Complexity vs. Practical Efficiency

The basic general algorithms of computational geometry have a double exponential worst case complexity. More precisely, if d is the maximal degree of the input polynomials and n the number of variables, their complexity is at most $d^{2^{cn}}$ for some constant c, and, for some inputs, the complexity is at least $d^{2^{c'n}}$ for another constant c′.

During the last 20 years of 20th century, various algorithms have been introduced to solve specific subproblems with a better complexity. Most of these algorithms have a complexity $d^{O(n^2)}$.

Among these algorithms which solve a sub problem of the problems solved by Gröbner bases, one may cite *testing if an affine variety is empty* and *solving nonhomogeneous polynomial systems which have a finite number of solutions.* Such algorithms are rarely implemented because, on most entries Faugère's F4 and F5 algorithms have a better practical efficiency and probably a similar or better complexity (*probably* because the evaluation of the complexity of Gröbner basis algorithms on a particular class of entries is a difficult task which has been done only in a few special cases).

The main algorithms of real algebraic geometry which solve a problem solved by CAD are related to the topology of semi-algebraic sets. One may cite *counting the number of connected components, testing if two points are in the same components* or *computing a Whitney stratification of a real algebraic set*. They have a complexity of $d^{O(n^2)}$, but the constant involved by O notation is so high that using them to solve any nontrivial problem effectively solved by CAD, is impossible even if one could use all the existing computing power in the world. Therefore, these algorithms have never been implemented and this is an active research area to search for algorithms with have together a good asymptotic complexity and a good practical efficiency.

Abstract Modern Viewpoint

The modern approaches to algebraic geometry redefine and effectively extend the range of basic objects in various levels of generality to schemes, formal schemes, ind-schemes, algebraic spaces, algebraic stacks and so on. The need for this arises already from the useful ideas within theory of varieties, e.g. the formal functions of Zariski can be accommodated by introducing nilpotent elements in structure rings; considering spaces of loops and arcs, constructing quotients by group actions and developing formal grounds for natural intersection theory and deformation theory lead to some of the further extensions.

Most remarkably, in late 1950s, algebraic varieties were subsumed into Alexander Grothendieck's concept of a scheme. Their local objects are affine schemes or prime spectra which are locally ringed spaces which form a category which is antiequivalent to the category of commutative unital rings, extending the duality between the category of affine algebraic varieties over a field k, and the category of finitely generated reduced k-algebras. The gluing is along Zariski topology; one can glue within the category of locally ringed spaces, but also, using the Yoneda embedding, within the more abstract category of presheaves of sets over the category of affine schemes. The Zariski topology in the set theoretic sense is then replaced by a Grothendieck topology. Grothendieck introduced Grothendieck topologies having in mind more exotic but geometrically finer and more sensitive examples than the crude Zariski topology, namely the étale topology, and the two flat Grothendieck topologies: fppf and fpqc; nowadays some other examples became prominent including Nisnevich topology. Sheaves can be furthermore generalized to stacks in the sense of Grothendieck, usually with some additional representability conditions leading to Artin stacks and, even finer, Deligne-Mumford stacks, both often called algebraic stacks.

Sometimes other algebraic sites replace the category of affine schemes. For example, Nikolai Durov has introduced commutative algebraic monads as a generalization of local objects in a generalized algebraic geometry. Versions of a tropical geometry, of an absolute geometry over a field of one element and an algebraic analogue of Arakelov's geometry were realized in this setup.

Another formal generalization is possible to universal algebraic geometry in which every variety of algebras has its own algebraic geometry. The term *variety of algebras* should not be confused with *algebraic variety*.

The language of schemes, stacks and generalizations has proved to be a valuable way of dealing with geometric concepts and became cornerstones of modern algebraic geometry.

Algebraic stacks can be further generalized and for many practical questions like deformation theory and intersection theory, this is often the most natural approach. One can extend the Grothendieck site of affine schemes to a higher categorical site of derived affine schemes, by replacing the commutative rings with an infinity category of differential graded commutative algebras, or of simplicial commutative rings or a similar category with an appropriate variant of a Grothendieck topology. One can also replace presheaves of sets by presheaves of simplicial sets (or of infinity groupoids). Then, in presence of an appropriate homotopic machinery one can develop a notion of derived stack as such a presheaf on the infinity category of derived affine schemes, which is satisfying certain infinite categorical version of a sheaf axiom (and to be algebraic, inductively a sequence of representability conditions). Quillen model categories, Segal categories and quasicategories are some of the most often used tools to formalize this yielding the *derived algebraic geometry*, introduced by the school of Carlos Simpson, including Andre Hirschowitz, Bertrand Toën, Gabrielle Vezzosi, Michel Vaquié and others; and developed further by Jacob Lurie, Bertrand Toën, and Gabrielle Vezzosi. Another (noncommutative) version of derived algebraic geometry, using A-infinity categories has been developed from early 1990s by Maxim Kontsevich and followers.

Analytic Geometry

An analytic variety is defined locally as the set of common solutions of several equations involving analytic functions. It is analogous to the included concept of real or complex algebraic variety. Any complex manifold is an analytic variety. Since analytic varieties may have singular points, not all analytic varieties are manifolds.

Modern analytic geometry is essentially equivalent to real and complex algebraic geometry, as has been shown by Jean-Pierre Serre in his paper *GAGA*, the name of which is French for *Algebraic geometry and analytic geometry*. Nevertheless, the two fields remain distinct, as the methods of proof are quite different and algebraic geometry includes also geometry in finite characteristic.

Applications

Algebraic geometry now finds applications in statistics, control theory, robotics, error-correcting codes, phylogenetics and geometric modelling. There are also connections to string theory, game theory, graph matchings, solitons and integer programming.

Algebraic Variety

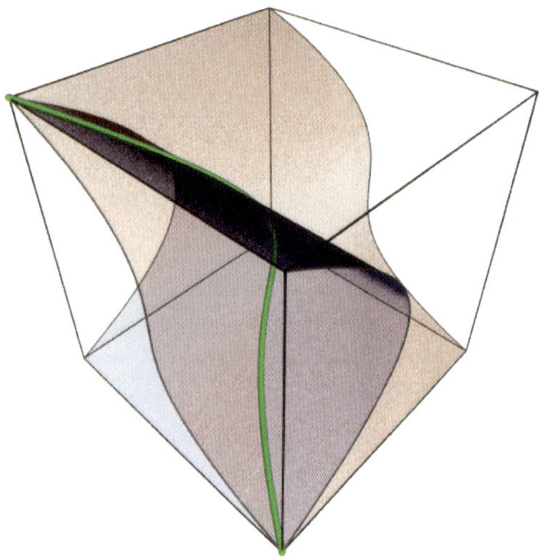

The twisted cubic is a projective algebraic variety.

Algebraic varieties are the central objects of study in algebraic geometry. Classically, an algebraic variety is defined as the set of solutions of a system of polynomial equations over the real or complex numbers. Modern definitions generalize this concept in several different ways, while attempting to preserve the geometric intuition behind the original definition.

Conventions regarding the definition of an algebraic variety differ slightly. For example, some definitions require an algebraic variety to be irreducible, which means that it is not the union of two smaller sets that are closed in the Zariski topology. Under this definition, non-irreducible algebraic varieties are called algebraic sets. Other conventions do not require irreducibility.

The concept of an algebraic variety is similar to that of an analytic manifold. An important difference is that an algebraic variety may have singular points, while a manifold cannot.

The fundamental theorem of algebra establishes a link between algebra and geometry by showing that a monic polynomial (an algebraic object) in one variable with complex number coefficients is determined by the set of its roots (a geometric object) in the complex plane. Generalizing this result, Hilbert's Nullstellensatz provides a fundamental correspondence between ideals of polynomial rings and algebraic sets. Using the Nullstellensatz and related results, mathematicians have established a strong correspondence between questions on algebraic sets and questions of ring theory. This correspondence is a defining feature of algebraic geometry.

Overview and Definitions

An *affine variety* over an algebraically closed field is conceptually the easiest type of variety to define, which will be done in this section. Next, one can define projective and quasi-projective varieties in a similar way. The most general definition of a variety is obtained by patching together smaller quasi-projective varieties. It is not obvious that one can construct genuinely new examples of varieties in this way, but Nagata gave an example of such a new variety in the 1950s.

Affine Varieties

For an algebraically closed field k and a natural number n, let A^n be affine n-space over k. The polynomials f in the ring $k[x_1, ..., x_n]$ can be viewed as k-valued functions on A^n by evaluating f at the points in A^n, i.e. by choosing values in k for each x_i. For each set S of polynomials in $k[x_1, ..., x_n]$, define the zero-locus $Z(S)$ to be the set of points in A^n on which the functions in S simultaneously vanish, that is to say

$$Z(S) = \{x \in \mathbf{A}^n \mid f(x) = 0 \text{ for all } f \in S\}.$$

A subset V of A^n is called an affine algebraic set if $V = Z(S)$ for some S. A nonempty affine algebraic set V is called irreducible if it cannot be written as the union of two proper algebraic subsets. An irreducible affine algebraic set is also called an affine variety. (Many authors use the phrase *affine variety* to refer to any affine algebraic set, irreducible or not)

Affine varieties can be given a natural topology by declaring the closed sets to be precisely the affine algebraic sets. This topology is called the Zariski topology.

Given a subset V of A^n, we define $I(V)$ to be the ideal of all polynomial functions vanishing on V:

$$I(V) = \{f \in k[x_1, \ldots, x_n] \mid f(x) = 0 \text{ for all } x \in V\}.$$

For any affine algebraic set V, the coordinate ring or structure ring of V is the quotient of the polynomial ring by this ideal.

Projective Varieties and Quasi-projective Varieties

Let k be an algebraically closed field and let P^n be the projective n-space over k. Let f in $k[x_0, ..., x_n]$ be a homogeneous polynomial of degree d. It is not well-defined to evaluate f on points in P^n in homogeneous coordinates. However, because f is homogeneous, meaning that $f(\lambda x_0, ..., \lambda x_n) = \lambda^d f(x_0, ..., x_n)$, it *does* make sense to ask whether f vanishes at a point $[x_0 : ... : x_n]$. For each set S of homogeneous polynomials, define the zero-locus of S to be the set of points in P^n on which the functions in S vanish:

$$Z(S) = \{x \in \mathbf{P}^n \mid f(x) = 0 \text{ for all } f \in S\}.$$

A subset V of P^n is called a projective algebraic set if $V = Z(S)$ for some S. An irreducible projective algebraic set is called a projective variety.

Projective varieties are also equipped with the Zariski topology by declaring all algebraic sets to be closed.

Given a subset V of P^n, let $I(V)$ be the ideal generated by all homogeneous polynomials vanishing on V. For any projective algebraic set V, the coordinate ring of V is the quotient of the polynomial ring by this ideal.

A quasi-projective variety is a Zariski open subset of a projective variety. Notice that every affine variety is quasi-projective. Notice also that the complement of an algebraic set in an affine variety is a quasi-projective variety; in the context of affine varieties, such a quasi-projective variety is usually not called a variety but a constructible set.

Abstract Varieties

In classical algebraic geometry, all varieties were by definition quasiprojective varieties, meaning that they were open subvarieties of closed subvarieties of projective space. For example, in Chapter 1 of Hartshorne a *variety* over an algebraically closed field is defined to be a quasi-projective variety, but from Chapter 2 onwards, the term variety (also called an abstract variety) refers to a more general object, which locally is a quasi-projective variety, but when viewed as a whole is not necessarily quasi-projective; i.e. it might not have an embedding into projective space. So classically the definition of an algebraic variety required an embedding into projective space, and this embedding was used to define the topology on the variety and the regular functions on the variety. The disadvantage of such a definition is that not all varieties come with natural embeddings into projective space. For example, under this definition, the product $P^1 \times P^1$ is not a variety until it is embedded into the projective space; this is usually done by the Segre embedding. However, any variety that admits one embedding into projective space admits many others by composing the embedding with the Veronese embedding. Consequently, many notions that should be intrinsic, such as the concept of a regular function, are not obviously so.

The earliest successful attempt to define an algebraic variety abstractly, without an embedding, was made by André Weil. In his *Foundations of Algebraic Geometry*, Weil defined an abstract algebraic variety using valuations. Claude Chevalley made a definition of a scheme, which served a similar purpose, but was more general. However, Alexander Grothendieck's definition of a scheme is more general still and has received the most widespread acceptance. In Grothendieck's language, an abstract algebraic variety is usually defined to be an integral, separated scheme of finite type over an algebraically closed field, although some authors drop the irreducibility or the reducedness or the separateness condition or allow the underlying field to be not algebraically closed. Classical algebraic varieties are the quasiprojective integral separated finite type schemes over an algebraically closed field.

Existence of Non-quasiprojective Abstract Algebraic Varieties

One of the earliest examples of a non-quasiprojective algebraic variety were given by Nagata. Nagata's example was not complete (the analog of compactness), but soon afterwards he found an algebraic surface that was complete and non-projective. Since then other examples have been found.

Examples

Subvariety

A subvariety is a subset of a variety that is itself a variety (with respect to the structure induced from the ambient variety). For example, every open subset of a variety is a variety.

Hilbert's Nullstellensatz says that closed subvarieties of an affine or projective variety are in one-to-one correspondence with the prime ideals or homogeneous prime ideals of the coordinate ring of the variety.

Affine Variety

Example 1

Let k = C, and A^2 be the two-dimensional affine space over C. Polynomials in the ring C[x, y] can be viewed as complex valued functions on A^2 by evaluating at the points in A^2. Let subset S of C[x, y] contain a single element f (x, y):

$$f(x, y) = x + y - 1.$$

The zero-locus of f (x, y) is the set of points in A^2 on which this function vanishes: it is the set of all pairs of complex numbers (x, y) such that $y = 1 - x$. This is called a line in the affine plane. (In the classical topology coming from the topology on the complex numbers, a complex line is a real manifold of dimension two.) This is the set $Z(f)$:

$$Z(f) = \{(x, 1-x) \in \mathbf{C}^2\}.$$

Thus the subset $V = Z(f)$ of A^2 is an algebraic set. The set V is not empty. It is irreducible, as it cannot be written as the union of two proper algebraic subsets. Thus it is an affine algebraic variety.

Example 2

Let k = C, and A^2 be the two-dimensional affine space over C. Polynomials in the ring C[x, y] can be viewed as complex valued functions on A^2 by evaluating at the points in A^2. Let subset S of C[x, y] contain a single element $g(x, y)$:

$$g(x, y) = x^2 + y^2 - 1.$$

The zero-locus of $g(x, y)$ is the set of points in A^2 on which this function vanishes, that is the set of points (x,y) such that $x^2 + y^2 = 1$. As $g(x, y)$ is an absolutely irreducible polynomial, this is an algebraic variety. The set of its real points (that is the points for which x and y are real numbers), is known as the unit circle; this name is also often given to the whole variety.

Example 3

The following example is neither a hypersurface, nor a linear space, nor a single point. Let A^3 be the three-dimensional affine space over C. The set of points (x, x^2, x^3) for x in C is an algebraic variety, and more precisely an algebraic curve that is not contained in any plane. It is the twisted cubic shown in the above figure. It may be defined by the equations

$$y - x^2 = 0$$

$$z - x^3 = 0$$

The irreducibility of this algebraic set needs a proof. One approach in this case is to check that the projection $(x, y, z) \rightarrow (x, y)$ is injective on the set of the solutions and that its image is an irreducible plane curve.

For more difficult examples, a similar proof may always be given, but may imply a difficult computation: first a Gröbner basis computation to compute the dimension, followed by a random linear change of variables (not always needed); then a Gröbner basis computation for another monomial ordering to compute the projection and to prove that it is generically injective and that its image is a hypersurface, and finally a polynomial factorization to prove the irreducibility of the image.

Projective Variety

A projective variety is a closed subvariety of a projective space. That is, it is the zero locus of a set of homogeneous polynomials that generate a prime ideal.

Example 1

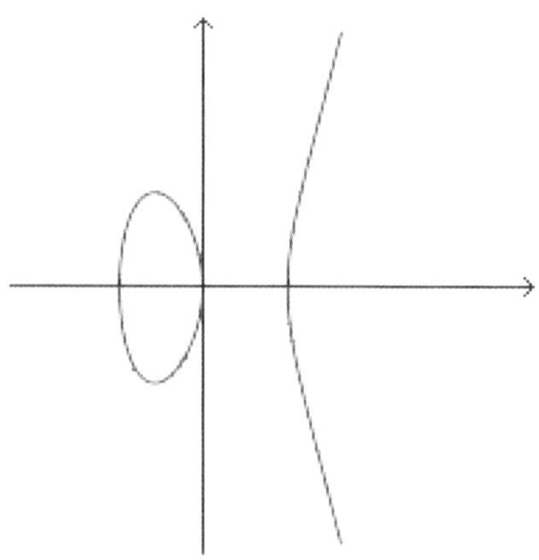

The affine plane curve $y^2 = x^3 - x$. The corresponding projective curve is called an elliptic curve.

A plane projective curve is the zero locus of an irreducible homogeneous polynomial in three indeterminates. The projective line P^1 is an example of a projective curve; it can be viewed as the curve in the projective plane $P^2 = \{[x,y,z]\}$ defined by $x = 0$. For another example, first consider the affine cubic curve

$$y^2 = x^3 - x.$$

in the 2-dimensional affine space (over a field of characteristic not two). It has the associated cubic homogeneous polynomial equation:

$$y^2 z = x^3 - xz^2,$$

which defines a curve in P^2 called an elliptic curve. The curve has genus one (genus formula); in particular, it is not isomorphic to the projective line P^1, which has genus zero. Using genus to distinguish curves is very basic: in fact, the genus is the first invariant one uses to classify curves.

Example 2

Let V be a finite-dimensional vector space. The Grassmannian variety $G_n(V)$ is the set of all n-dimensional subspaces of V. It is a projective variety: it is embedded into a projective space via the Plücker embedding:

$$G_n(V) \hookrightarrow \mathbf{P}(\wedge^n V), \langle b_1, \ldots, b_n \rangle \mapsto [b_1 \wedge \cdots \wedge b_n]$$

where b_i are any set of linearly independent vectors in V, $\wedge^n V$ is the n-th exterior power of V and the bracket $[w]$ means the line spanned by the nonzero vector w.

The Grassmannian variety comes with a natural vector bundle (or locally free sheaf in other terminology) called the tautological bundle, which is important in the study of characteristic classes such as Chern classes.

Non-affine and Non-projective Example

An algebraic variety can be neither affine nor projective. To give an example, let $X = P^1 \times A^1$ and p: $X \to A^1$ the projection. It is an algebraic variety since it is a product of varieties. It is not affine since P^1 is a closed subvariety of X (as the zero locus of p), but an affine variety cannot contain a projective variety of positive dimension as a closed subvariety. It is not projective either, since there is a nonconstant regular function on X; namely, p.

Another example of a non-affine non-projective variety is $X = A^2 - (0, 0)$ (cf. morphism of varieties#Examples.)

Basic Results

- An affine algebraic set V is a variety if and only if $I(V)$ is a prime ideal; equivalently, V is a variety if and only if its coordinate ring is an integral domain.
- Every nonempty affine algebraic set may be written uniquely as a finite union of algebraic varieties (where none of the varieties in the decomposition is a subvariety of any other).
- The dimension of a variety may be defined in various equivalent ways.

Isomorphism of Algebraic Varieties

Let V_1, V_2 be algebraic varieties. We say V_1 and V_2 are isomorphic, and write $V_1 \cong V_2$, if there are regular maps $\varphi : V_1 \to V_2$ and $\psi : V_2 \to V_1$ such that the compositions $\psi \circ \varphi$ and $\varphi \circ \psi$ are the identity maps on V_1 and V_2 respectively.

Discussion and Generalizations

The basic definitions and facts above enable one to do classical algebraic geometry. To be able to do more — for example, to deal with varieties over fields that are not algebraically closed — some foundational changes are required. The modern notion of a variety is considerably more abstract than the one above, though equivalent in the case of varieties over algebraically closed fields. An *abstract algebraic variety* is a particular kind of scheme; the generalization to schemes on the geometric side enables an extension of the correspondence described above to a wider class of rings. A scheme is a locally ringed space such that every point has a neighbourhood that, as a locally ringed space, is isomorphic to a spectrum of a ring. Basically, a variety over k is a scheme whose structure sheaf is a sheaf of k-algebras with the property that the rings R that occur above are all integral domains and are all finitely generated k-algebras, that is to say, they are quotients of polynomial algebras by prime ideals.

This definition works over any field k. It allows you to glue affine varieties (along common open sets) without worrying whether the resulting object can be put into some projective space. This also leads to difficulties since one can introduce somewhat pathological objects, e.g. an affine line with zero doubled. Such objects are usually not considered varieties, and are eliminated by requiring the schemes underlying a variety to be *separated*. (Strictly speaking, there is also a third condition, namely, that one needs only finitely many affine patches in the definition above.)

Some modern researchers also remove the restriction on a variety having integral domain affine charts, and when speaking of a variety only require that the affine charts have trivial nilradical.

A complete variety is a variety such that any map from an open subset of a nonsingular curve into it can be extended uniquely to the whole curve. Every projective variety is complete, but not vice versa.

These varieties have been called "varieties in the sense of Serre", since Serre's foundational paper FAC on sheaf cohomology was written for them. They remain typical objects to start studying in algebraic geometry, even if more general objects are also used in an auxiliary way.

One way that leads to generalizations is to allow reducible algebraic sets (and fields k that aren't algebraically closed), so the rings R may not be integral domains. A more significant modification is to allow nilpotents in the sheaf of rings, that is, rings which are not reduced. This is one of several generalizations of classical algebraic geometry that are built into Grothendieck's theory of schemes.

Allowing nilpotent elements in rings is related to keeping track of "multiplicities" in algebraic geometry. For example, the closed subscheme of the affine line defined by $x^2 = 0$ is different from the subscheme defined by $x = 0$ (the origin). More generally, the fiber of a morphism of schemes $X \to Y$ at a point of Y may be non-reduced, even if X and Y are reduced. Geometrically, this says that fibers of good mappings may have nontrivial "infinitesimal" structure.

There are further generalizations called algebraic spaces and stacks.

Algebraic Manifolds

An algebraic manifold is an algebraic variety that is also an m-dimensional manifold, and hence every sufficiently small local patch is isomorphic to k^m. Equivalently, the variety is smooth (free from singular points). When k is the real numbers, R, algebraic manifolds are called Nash manifolds. Algebraic manifolds can be defined as the zero set of a finite collection of analytic algebraic functions. Projective algebraic manifolds are an equivalent definition for projective varieties. The Riemann sphere is one example.

Affine Variety

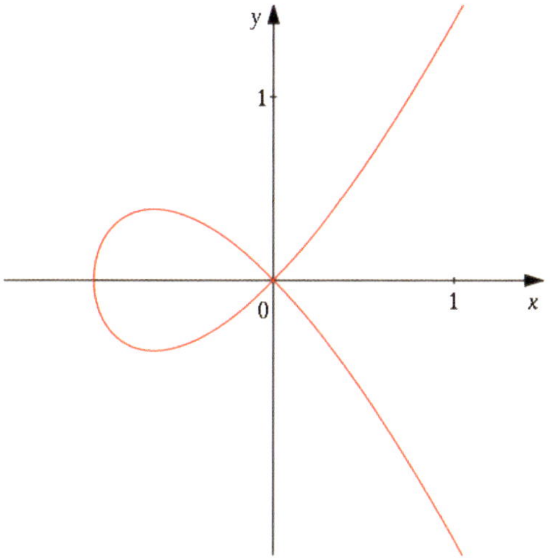

A cubic plane curve given by $y^2 = x^2(x+1)$

In algebraic geometry, an affine variety over an algebraically closed field k is the zero-locus in the affine n-space k^n of some finite family of polynomials of n variables with coefficients in k that generate a prime ideal. If the condition of generating a prime ideal is removed, such a set is called an (affine) algebraic set. A Zariski open subvariety of an affine variety is called a quasi-affine variety.

If X is an affine variety defined by a prime ideal I, then the quotient ring

$$k[x_1,\ldots,x_n]/I$$

is called the coordinate ring of X. This ring is precisely the set of all regular functions on X; in other words, it is the space of global sections of the structure sheaf of X. A theorem of Serre gives a cohomological characterization of an affine variety; it says an algebraic variety is affine if and only if

$$H^i(X,F)=0$$

for any $i>0$ and any quasi-coherent sheaf F on X. (cf. Cartan's theorem B.) This makes the cohomological study of an affine variety non-existent, in a sharp contrast to the projective case in which cohomology groups of line bundles are of central interest.

An affine variety plays a role of a local chart for algebraic varieties; that is to say, general algebraic varieties such as projective varieties are obtained by gluing affine varieties. Linear structures that are attached to varieties are also (trivially) affine varieties; e.g., tangent spaces, fibers of algebraic vector bundles.

An affine variety is, up to an equivalence of categories a special case of an affine scheme, which is precisely the spectrum of a ring. In complex geometry, an affine variety is an analog of a Stein manifold.

Introduction

The most concrete point of view to describe an affine algebraic variety is that it is the set of solutions in an algebraically closed field k of a system of polynomial equations with coefficients in k. More precisely, if $f_1,\ldots,f_k$ are polynomials with coefficients in k, they define an affine variety (or affine algebraic set)

$$V(f_1,\ldots,f_k)=\left\{(a_1,\ldots,a_n)\in k^n \mid f_1(a_1,\ldots,a_n)=\ldots=f_k(a_1,\ldots,a_n)=0\right\}.$$

By Hilbert's Nullstellensatz, the points of the variety are in one-to-one correspondence with the maximal ideals of its *coordinate ring*, the k-algebra $R=k[x_1,\ldots,x_n]/\langle f_1,\ldots,f_k\rangle$, through the map $(a_1,\ldots,a_n)\mapsto\langle \overline{x_1-a_1},\ldots,\overline{x_n-a_n}\rangle$, where $\overline{x_i-a_i}$ denotes the image in the quotient algebra R of the polynomial x_i-a_i. In scheme theory, this correspondence has been extended to prime ideals to define the affine scheme $\mathrm{Spec}(R)$, which may be identified to the variety, through an equivalence of categories.

The elements of the coordinate ring R are also called the *regular functions* or the *polynomial functions* on the variety. They form the *ring of the regular functions* on the variety, or, simply, the *ring of the variety*. In fact an element $\overline{f}\in R$ is the image of a polynomial $f\in k[x_1,\ldots,x_n]$, which defines a function from k^n into k; The restriction of f to the variety does not depend on the choice of f among the polynomials mapped on $\overline{f}$ by the quotient.

The dimension of a variety is an integer associated to every variety, and even to every algebraic set, whose importance relies on the large number of its equivalent definitions.

Structure Sheaf

Equipped with the structure sheaf described below, an affine variety is a locally ringed space.

Given an affine variety X with coordinate ring A, we define the sheaf of k-algebras $\mathcal{O}_X$ by letting $\mathcal{O}_X(U)=\Gamma(U,\mathcal{O}_X)$ be the ring of regular functions on U.

We let $D(f) = \{ x \mid f(x) \neq 0 \}$ for each f in A. They form a base for the topology of X and so $\mathcal{O}_X$ is determined by its values on the open sets $D(f)$.

The key fact, which relies on Hilbert nullstellensatz in the essential way, is the following:

Claim — $\Gamma(D(f),\mathcal{O}_X)=A[f^{-1}]$ for any f in A.

Proof: The inclusion $\supset$ is clear. For the opposite, let g be in the left-hand side and $J=\{h\in A \mid hg\in A\}$, which is an ideal. If x is in $D(f)$, then, since g is regular near x, there is some open affine neighborhood $D(h)$ of x such that $g\in k[D(h)]=A[h^{-1}]$; that is, $h^m g$ is in A and thus x is not in $V(J)$. In other words, $V(J)\subset\{x \mid f(x)=0\}$ and thus the Hilbert nullstellensatz implies f is in the radical of J; i.e., $f^n g\in A$.

The claim, first of all, implies that X is a "locally ringed" space since

$$\mathcal{O}_{X,x}=\varinjlim_{f(x)\neq 0} A[f^{-1}]=A_{\mathfrak{m}_x}$$

where $\mathfrak{m}_x=\{f\in A \mid f(x)=0\}$. Secondly, the claim implies that $\mathcal{O}_X$ is a sheaf; indeed, it says if a function is regular (pointwise) on $D(f)$, then it must be in the coordinate ring of $D(f)$; that is, "regular-ness" can be patched together.

Hence, $(X, \mathcal{O}_X)$ is a locally ringed space.

Examples

- The complement of a hypersurface of an affine variety X; i.e., $X - \{f = 0\}$ for some regular function f on X is affine; its coordinate ring is the localization $k[X][f^{-1}]$. In particular, $A^1 - 0$ (the affine line with the origin removed) is affine.
- Every closed subvariety of the affine space $\mathbf{A}^n$ of codimension one is defined by a prime ideal of the polynomial ring of height one, which is principal; thus, they are hypersurfaces (i.e., defined by a single polynomial.)
- $C^2 - 0$ is an open subset of the affine variety that is not affine; cf. Hartogs' extension theorem
- The normalization of an irreducible affine variety is affine; the coordinate ring of the normalization is the integral closure of the coordinate ring of the variety. (It turns out the normalization of a projective variety is a projective variety.)

Tangent Space

Tangent spaces may be defined just as in calculus. Let $X = \operatorname{spec} A$, $A = k[x_1, \ldots, x_n]/(f_1, \ldots, f_r)$ be the affine variety. Then the affine subvariety of k^n defined by the linear equations

$$\sum_{i=1}^{n} \frac{\partial f_j}{\partial x_i}(a_1, \ldots, a_n)(x_i - a_i) = 0, \quad j = 1, \ldots, r$$

is called the tangent space at $x = (a_1, \ldots, a_n)$. (A more intrinsic definition is given by Zariski tangent space.) If the tangent space at x and the variety X have the same dimension, the point x is said to be smooth; otherwise, singular.

The important difference from calculus is that the inverse function theorem fails. To alleviate this problem, one has to consider the étale topology instead of the Zariski topology. (cf. Milne, Étale)

Projective Variety

In algebraic geometry, a projective variety over an algebraically closed field k is a subset of some projective n-space P^n over k that is the zero-locus of some finite family of homogeneous polynomials of $n + 1$ variables with coefficients in k, that generate a prime ideal, the defining ideal of the variety. Equivalently, an algebraic variety is projective if it can be embedded as a Zariski closed subvariety of P^n.

A projective variety is a projective curve if its dimension is one; it is a projective surface if its dimension is two; it is a projective hypersurface if its dimension is one less than the dimension of the containing projective space; in this case it is the set of zeros of a single homogeneous polynomial.

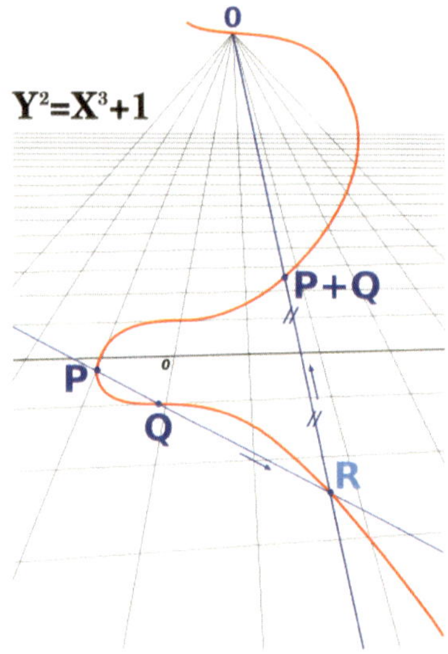

An elliptic curve is a smooth projective curve of genus one.

If X is a projective variety defined by a homogeneous prime ideal I, then the quotient ring

$$k[x_0,\dots,x_n]/I$$

is called the homogeneous coordinate ring of X. Basic invariants of X such as the degree and the dimension can be read off the Hilbert polynomial of this graded ring.

Projective varieties arise in many ways. They are complete, which roughly can be expressed by saying that there are no points "missing". The converse is not true in general, but Chow's lemma describes the close relation of these two notions. Showing that a variety is projective is done by studying line bundles or divisors on X.

A salient feature of projective varieties are the finiteness constraints on sheaf cohomology. For smooth projective varieties, Serre duality can be viewed as an analog of Poincaré duality. It also leads to the Riemann-Roch theorem for projective curves, i.e., projective varieties of dimension 1. The theory of projective curves is particularly rich, including a classification by the genus of the curve. The classification program for higher-dimensional projective varieties naturally leads to the construction of moduli of projective varieties. Hilbert schemes parametrize closed subschemes of P^n with prescribed Hilbert polynomial. Hilbert schemes, of which Grassmannians are special cases, are also projective schemes in their own right. Geometric invariant theory offers another approach. The classical approaches include the Teichmüller space and Chow varieties.

A particularly rich theory, reaching back to the classics, is available for complex projective varieties, i.e., when the polynomials defining X have complex coefficients. Broadly, the GAGA principle says that the geometry of projective complex analytic spaces (or manifolds) is equivalent to the geometry of projective complex varieties. For example, the theory of holomorphic vector bundles (more generally coherent analytic sheaves) on X coincide with that of algebraic vector bundles. Chow's theorem says that a subset of projective space is the zero-locus of a family of holomorphic functions if and only if it is the zero-locus of homogeneous polynomials. The combination of analytic and algebraic methods for complex projective varieties lead to areas such as Hodge theory.

Variety and Scheme Structure

Variety Structure

Let k be an algebraically closed field. The basis of the definition of projective varieties is projective space P^n, which can be defined in different, but equivalent ways:

- as the set of all lines through the origin in k^{n+1} (i.e., one-dimensional sub-vector spaces of k^{n+1})
- as the set of tuples $(x_0,\ldots,x_n)\in k^{n+1}$, modulo the equivalence relation

  $$(x_0,\ldots,x_n)\sim\lambda(x_0,\ldots,x_n)$$

  for any $\lambda\in k\setminus\{0\}$. The equivalence class of such a tuple is denoted by

  $$[x_0:\cdots:x_n]$$

  and referred to as a homogeneous coordinate.

A *projective variety* is, by definition, a closed subvariety of P^n, where closed refers to the Zariski topology. In general, closed subsets of the Zariski topology are defined to be the zero-locus of polynomial functions. Given a polynomial $f\in k[x_0,\ldots,x_n]$, the condition

$$f([x_0:\cdots:x_n])=0$$

does not make sense for arbitrary polynomials, but only if f is homogeneous, i.e., the total degree of all the monomials (whose sum is f) is the same. In this case, the vanishing of

$$f(\lambda x_0,\ldots,\lambda x_n)=\lambda^{\deg f}f(x_0,\ldots,x_n)$$

is independent of the choice of $\lambda(\neq 0)$.

Therefore, projective varieties arise from homogeneous prime ideals I of $k[x_0,\ldots,x_n]$, and setting

$$X=\{[x_0:\cdots:x_n]\in\mathbf{P}^n, f([x_0:\cdots:x_n])=0\text{ for all }f\in I\}.$$

Moreover, the projective variety X is an algebraic variety, meaning that it is covered by open affine subvarieties and satisfies the separation axiom. Thus, the local study of X (e.g., singularity) reduces to that of an affine variety. The explicit structure is as follows. The projective space P^n is covered by the standard open affine charts

$$U_i=\{[x_0:\cdots:x_n], x_i\neq 0\},$$

which themselves are affine n-spaces with the coordinate ring $k[y_1^{(i)},\ldots,y_n^{(i)}], y_j^{(i)}=x_j/x_i$. Say $i=0$ for the notational simplicity and drop the superscript (0). Then $X\cap U_0$ is a closed subvariety of $U_0\simeq\mathbb{A}^n$ defined by the ideal of $k[y_1,\ldots,y_n]$ generated by

$$f(1,y_1,\ldots,y_n)$$

for all f in I. Thus, X is an algebraic variety covered by $(n+1)$ open affine charts $X \cap U_i$.

Note that X is the closure of the affine variety $X \cap U_0$ in $\mathbb{P}^n$. Conversely, starting from some closed (affine) variety $V \subset U_0 \simeq \mathbb{A}^n$, the closure of V in $\mathbb{P}^n$ is the projective variety called the projective completion of V. If $I \subset k[y_1,\ldots,y_n]$ defines V, then the defining ideal of this closure is the homogeneous ideal of $k[x_0,\ldots,x_n]$ generated by

$$x_0^{\deg(f)} f(x_1 / x_0,\ldots,x_n / x_0)$$

for all f in I.

For example, if V is an affine curve given by, say, $y^2 = x^3 + ax + b$ in the affine plane, then its projective completion in the projective plane is given by $y^2 z = x^3 + axz^2 + bz^3$.

Projective Schemes

For various applications, it is necessary to consider more general algebro-geometric objects than projective varieties, namely projective schemes. The first step towards projective schemes is to endow projective space with a scheme structure, in a way refining the above description of projective space as an algebraic variety, i.e., $P^n(k)$ is a scheme which it is a union of $(n + 1)$ copies of the affine n-space k^n. More generally, projective space over a ring A is the union of the affine schemes

$$U_i = \operatorname{Spec} A[x_1 / x_i,\ldots,x_n / x_i], \quad 0 \le i \le n,$$

in such a way the variables match up as expected. The set of closed points of $\mathbf{P}_k^n$, for algebraically closed fields k, is then the projective space $P^n(k)$ in the usual sense.

An equivalent but streamlined construction is given by the Proj construction, which is an analog of the spectrum of a ring, denoted "Spec", which defines an affine scheme. For example, if A is a ring, then

$$\mathbf{P}_A^n = \operatorname{Proj} A[x_0,\ldots,x_n].$$

If R is a quotient of $k[x_0,\ldots,x_n]$ by a homogeneous ideal I, then the canonical surjection induces the closed immersion

$$\operatorname{Proj} R \hookrightarrow \mathbf{P}_k^n.$$

Compared to projective varieties, the condition that the ideal I be a prime ideal was dropped. This leads to a much more flexible notion: on the one hand the topological space $X = \operatorname{Proj} R$ may have multiple irreducible components. Moreover, there may be nilpotent functions on X.

Closed subschemes of $\mathbf{P}_k^n$ correspond bijectively to the homogeneous ideals I of $k[x_0,\ldots,x_n]$ that are saturated; i.e., $I : (x_0,\ldots,x_n) = I$. This fact may be considered as a refined version of projective Nullstellensatz.

We can give a coordinate-free analog of the above. Namely, given a finite-dimensional vector space V over k, we let

$$\mathbf{P}(V) = \operatorname{Proj} k[V]$$

where $k[V] = \operatorname{Sym}(V^*)$ is the symmetric algebra of V^*. It is the projectivization of *V*; i.e., it parametrizes lines in *V*. There is a canonical surjective map $\pi : V - 0 \to \mathbf{P}(V)$, which is defined using the chart described above. One important use of the construction is this (cf., Duality and linear system). A divisor *D* on a projective variety *X* corresponds to a line bundle *L*. One then set

$$|D| = \mathbf{P}(\Gamma(X, L));$$

it is called the complete linear system of *D*.

Projective space over any scheme *S* can be defined as a fiber product of schemes

$$\mathbf{P}_S^n = \mathbf{P}_{\mathbf{Z}}^n \times_{\operatorname{Spec} \mathbf{Z}} S.$$

If $\mathcal{O}(1)$ is the twisting sheaf of Serre on $\mathbf{P}_{\mathbf{Z}}^n$, we let $\mathcal{O}(1)$ denote the pullback of $\mathcal{O}(1)$ to $\mathbf{P}_S^n$; that is, $\mathcal{O}(1) = g^*(\mathcal{O}(1))$ for the canonical map $g : \mathbf{P}_S^n \to \mathbf{P}_{\mathbf{Z}}^n$.

A scheme $X \to S$ is called projective over *S* if it factors as a closed immersion

$$X \to \mathbf{P}_S^n$$

followed by the projection to *S*.

A line bundle (or invertible sheaf) $\mathcal{L}$ on a scheme *X* over *S* is said to be very ample relative to *S* if there is an immersion (i.e., an open immersion followed by a closed immersion)

$$i : X \to \mathbf{P}_S^n$$

for some *n* so that $\mathcal{O}(1)$ pullbacks to $\mathcal{L}$. Then a *S*-scheme *X* is projective if and only if it is proper and there exists a very ample sheaf on *X* relative to *S*. Indeed, if *X* is proper, then an immersion corresponding to the very ample line bundle is necessarily closed. Conversely, if *X* is projective, then the pullback of $\mathcal{O}(1)$ under the closed immersion of *X* into a projective space is very ample. That "projective" implies "proper" is deeper: the *main theorem of elimination theory*.

Relation to Complete Varieties

By definition, a variety is complete, if it is proper over *k*. The valuative criterion of properness expresses the intuition that in a proper variety, there are no points "missing".

There is a close relation between complete and projective varieties: on the one hand, projective space and therefore any projective variety is complete. The converse is not true in general. However:

- A smooth curve *C* is projective if and only if it is complete. This is proved by identifying *C* with the set of discrete valuation rings of the function field *k*(*C*) over *k*. This set has a natural Zariski topology called the Zariski–Riemann space.

- Chow's lemma states that for any complete variety X, there is a projective variety Z and a birational morphism $Z \to X$. (Moreover, through normalization, one can assume this projective variety is normal.)

Some properties of a projective variety follow from completeness. For example,

$$\Gamma(X, \mathcal{O}_X) = k$$

for any projective variety X over k. This fact is an algebraic analogue of Liouville's theorem (any holomorphic function on a connected compact complex manifold is constant). In fact, the similarity between complex analytic geometry and algebraic geometry on complex projective varieties goes much further than this, as is explained below.

Quasi-projective varieties are, by definition, those which are open subvarieties of projective varieties. This class of varieties includes affine varieties. Affine varieties are almost never complete (or projective). In fact, a projective subvariety of an affine variety must have dimension zero. This is because only the constants are globally regular functions on a projective variety.

Examples and Basic Invariants

By definition, any homogeneous ideal in a polynomial ring yields a projective scheme (required to be prime ideal to give a variety). In this sense, examples of projective varieties abound. The following list mentions various classes of projective varieties which are noteworthy since they have been studied particularly intensely. The important class of complex projective varieties, i.e., the case $k = \mathbf{C}$, is discussed further below.

The product of two projective spaces is projective. In fact, there is the explicit immersion (called Segre embedding)

$$\mathbf{P}^n \times \mathbf{P}^m \to \mathbf{P}^{(n+1)(m+1)-1}, (x_i, y_j) \mapsto x_i y_j.$$

As a consequence, the product of projective varieties over k is again projective. The Plücker embedding exhibits a Grassmannian as a projective variety. Flag varieties such as the quotient of the general linear group $GL_n(k)$ modulo the subgroup of upper triangular matrices, are also projective, which is an important fact in the theory of algebraic groups.

Homogeneous Coordinate Ring and Hilbert Polynomial

As the prime ideal P defining a projective variety X is homogeneous, the homogeneous coordinate ring

$$R = k[x_0, \ldots, x_n] / P$$

is a graded ring, i.e., can be expressed as the direct sum of its graded components:

$$R = \bigoplus_{n \in \mathbf{N}} R_n.$$

There exists a polynomial P such that $\dim R_n = P(n)$ for all sufficiently large n; it is called the Hilbert polynomial of X. It is a numerical invariant encoding some extrinsic geometry of X. The degree of P is the dimension r of X and its leading coefficient times r! is the degree of the variety X. The arithmetic genus of X is $(-1)^r (P(0) - 1)$ when X is smooth.

For example, the homogeneous coordinate ring of P^n is $k[x_0,\ldots,x_n]$ and its Hilbert polynomial is $P(z) = \binom{z+n}{n}$; its arithmetic genus is zero.

If the homogeneous coordinate ring R is an integrally closed domain, then the projective variety X is said to be projectively normal. Note, unlike normality, projective-normality depends on R, the embedding of X into a projective space. The normalization of a projective variety is projective; in fact, it's the Proj of the integral closure of some homogeneous coordinate ring of X.

Degree

Let $X \subset \mathbb{P}^N$ be a projective variety. There are at least two equivalent ways to define the degree of X relative to its embedding. The first way is to define it as the cardinality of the finite set

$$\#(X \cap H_1 \cap \cdots \cap H_d)$$

where d is the dimension of X and H_i's are hyperplanes in "general positions". This definition corresponds to an intuitive idea of a degree. Indeed, if X is a hypersurface, then the degree of X is the degree of the homogeneous polynomial defining X. The "general positions" can be made precise, for example, by intersection theory; one requires that the intersection is proper and that the multiplicities of irreducible components are all one.

The other definition, which is mentioned in the previous section, is that the degree of X is the leading coefficient of the Hilbert polynomial of X times (dim X)!. Geometrically, this definition means that the degree of X is the multiplicity of the vertex of the affine cone over X.

Let $V_1,\ldots,V_r \subset \mathbb{P}^N$ be closed subschemes of pure dimensions that intersect properly (they are in general position). If m_i denotes the multiplicity of an irreducible component Z_i in the intersection (i.e., intersection multiplicity), then the generalization of Bézout's theorem says:

$$\sum_1^s m_i \deg Z_i = \prod_1^r \deg V_i.$$

The intersection multiplicity m_i can be defined as the coefficient of Z_i in the intersection product $V_1 \cdot \ldots \cdot V_r$ in the Chow ring of $\mathbb{P}^N$.

In particular, if $H \subset \mathbb{P}^N$ is a hypersurface not containing X, then

$$\sum_1^s m_i \deg Z_i = \deg(X)\deg(H)$$

where Z_i's are the irreducible components of the scheme-theoretic intersection of X and H with multiplicity (length of the local ring) m_i.

The Ring of Sections

Let X be a projective variety and L a line bundle on it. Then the graded ring

$$R(X,L) = \bigoplus_{n=0}^{\infty} H^0(X, L^{\otimes n})$$

is called the ring of sections of L. If L is ample, then Proj of this ring is X. Moreover, if X is normal and L is very ample, then $R(X,L)$ is the integral closure of the homogeneous coordinate ring of X determined by L; i.e., $X \hookrightarrow \mathbb{P}^N$ so that $\mathcal{O}_{\mathbb{P}^N}(1)$ pulls-back to L.

For applications, it is useful to allow for divisors (or $\mathbb{Q}$-divisors) not just line bundles; assuming X is normal, the resulting ring is then called a generalized ring of sections. If K_X is a canonical divisor on X, then the generalized ring of sections

$$R(X, K_X)$$

is called the canonical ring of X. If the canonical ring is finitely generated, then Proj of the ring is called the canonical model of X. The canonical ring or model can then be used to define the Kodaira dimension of X.

Projective Curves

Projective schemes of dimension one are called *projective curves*. Much of the theory of projective curves is about smooth projective curves, since the singularities of curves can be resolved by normalization, which consists in taking locally the integral closure of the ring of regular functions. Smooth projective curves are isomorphic if and only if their function fields are isomorphic. The study of finite extensions of

$$\mathbb{F}_p(t),$$

or equivalently smooth projective curves over $\mathbb{F}_p$ is an important branch in algebraic number theory.

A smooth projective curve of genus one is called an elliptic curve. As a consequence of the Riemann-Roch theorem, such a curve can be embedded as a closed subvariety in P^2. In general, any (smooth) projective curve can be embedded in P^3. Conversely, any smooth closed curve in P^2 of degree three has genus one by the genus formula and is thus an elliptic curve.

A smooth complete curve of genus greater than or equal to two is called a hyperelliptic curve if there is a finite morphism $C \to \mathbf{P}^1$ of degree two.

Projective Hypersurfaces

Every irreducible closed subset of P^n of codimension one is a hypersurface; i.e., the zero set of some homogeneous irreducible polynomial.

Abelian Varieties

Another important invariant of a projective variety X is the Picard group $\text{Pic}(X)$ of X, the set of isomorphism classes of line bundles on X. It is isomorphic to $H^1(X,\mathcal{O}_X^*)$ and therefore an intrinsic notion (independent of embedding). For example, the Picard group of P^n is isomorphic to Z via the degree map. The kernel of $\deg : \text{Pic}(X) \to \mathbf{Z}$ is not only an abstract abelian group, but there is a variety called the Jacobian variety of X, Jac(X), whose points equal this group. The Jacobian of a (smooth) curve plays an important role in the study of the curve. For example, the Jacobian of an elliptic curve E is E itself. For a curve X of genus g, Jac(X) has dimension g.

Varieties, such as the Jacobian variety, which are complete and have a group structure are known as abelian varieties, in honor of Niels Abel. In marked contrast to affine algebraic groups such as $GL_n(k)$, such groups are always commutative, whence the name. Moreover, they admit an ample line bundle and are thus projective. On the other hand, an abelian scheme may not be projective. Examples of abelian varieties are elliptic curves, Jacobian varictics and K3 surfaces.

Projections

Let $E \subset \mathbb{P}^n$ be a linear subspace; i.e., $E = \{s_0 = s_1 = \ldots = s_r = 0\}$ for some linearly independent linear functional s_i's. Then the projection from ***E*** is the (well-defined) morphism

$$\phi : \mathbb{P}^n - E \to \mathbb{P}^r, x \mapsto [s_0(x) : \cdots : s_r(x)].$$

- The geometric description of this map is the following. We view $\mathbb{P}^r \subset \mathbb{P}^n$ so that it is disjoint from E. Then, for any $x \in \mathbb{P}^n - E$,

  $$\phi(x) = W_x \cap \mathbb{P}^r$$

  where we wrote W_x for the smallest linear space containing E and x (called the join of E and x.)
- $\phi^{-1}(\{y_i \neq 0\}) = \{s_i \neq 0\}$, where y_i are the homogeneous coordinates on $\mathbb{P}^r$.
- For any closed subscheme $Z \subset \mathbb{P}^n$ disjoint from E, the restriction

  $$\phi : Z \to \mathbb{P}^r$$

  is a finite morphism.

Projections can be used to cut down the dimension in which a projective variety is embedded, up to finite morphisms. Start with some projective variety $X \subset \mathbb{P}^n$. If $n > \dim X$, the projection from a point not on X gives $\phi : X \to \mathbb{P}^{n-1}$. Moreover, ϕ is a finite map to its image. Thus, iterating the procedure, one sees there is a finite map

$$X \to \mathbb{P}^d, d = \dim X .$$

This result is the projective analog of Noether's normalization lemma. (In fact, it yields a geometric proof of the normalization lemma.)

The same procedure can be used to show the following slightly more precise result: given a projective variety X over a perfect field, there is a finite birational morphism from X to a hypersurface H in $\mathbb{P}^{d+1}$. In particular, if X is normal, then it is the normalization of H.

Duality and Linear System

While a projective n-space $\mathbb{P}^n$ parameterizes the lines in an affine n-space, the dual of it parametrizes the hyperplanes on the projective space, as follows. Fix a field k. By $\check{\mathbb{P}}_k^n$, we mean a projective n-space

$$\check{\mathbb{P}}_k^n = \operatorname{Proj}(k[u_0, \ldots, u_n])$$

equipped with the construction:

$$f \mapsto H_f = \{\alpha_0 x_0 + \cdots + \alpha_n x_n = 0\}, \text{ a hyperplane on } \mathbb{P}_L^n$$

where $f : \operatorname{Spec} L \to \check{\mathbb{P}}_k^n$ is an L-point of $\check{\mathbb{P}}_k^n$ for a field extension L of k and $\alpha_i = f^*(u_i) \in L$.

For each L, the construction is a bijection between the set of L-points of $\check{\mathbb{P}}_k^n$ and the set of hyperplanes on $\mathbb{P}_L^n$. Because of this, the dual projective space $\check{\mathbb{P}}_k^n$ is said to be the moduli space of hyperplanes on .

A line in $\check{\mathbb{P}}_k^n$ is called a pencil: it is a family of hyperplanes on $\mathbb{P}_k^n$ parametrized by $\mathbb{P}_k^1$.

If V is a finite-dimensional vector space over k, then, for the same reason as above, $\mathbb{P}(V^*) = \operatorname{Proj}(\operatorname{Sym}(V))$ is the space of hyperplanes on $\mathbb{P}(V)$. An important case is when V consists of sections of a line bundle. Namely, let X be an algebraic variety, L a line bundle on X and $V \subset \Gamma(X, L)$ a vector subspace of finite positive dimension. Then there is

$$\varphi_V : X - B \to \mathbb{P}(V^*)$$

given by

$$x \mapsto H_x = \{s \in V \mid s(x) = 0\}$$

and called the map determined by the linear system V, where B, called the base locus, is the intersection of the divisors of zero of nonzero sections in V.

Cohomology of Coherent Sheaves

Let X be a projective scheme over a field (or, more generally over a Noetherian ring A). Cohomology of coherent sheaves $\mathcal{F}$ on X satisfies the following important theorems due to Serre:

1. $H^p(X, \mathcal{F})$ is a finite-dimensional k-vector space for any p.
2. There exists an integer n_0 (depending on $\mathcal{F}$; Castelnuovo–Mumford regularity) such that

$$H^p(X, \mathcal{F}(n)) = 0$$

for all $n \geq n_0$ and p > 0, where $\mathcal{F}(n) = \mathcal{F} \otimes \mathcal{O}(n)$ is the twisting with a power of a very ample line bundle $\mathcal{O}(1)$

These results are proven reducing to the case $X = \mathbf{P}^n$ using the isomorphism

$$H^p(X, \mathcal{F}) = H^p(\mathbf{P}^r, \mathcal{F}), p \geq 0$$

where in the right-hand side $\mathcal{F}$ is viewed as a sheaf on the projective space by extension by zero. The result then follows by a direct computation for $\mathcal{F} = \mathcal{O}_{\mathbf{P}^r}(n)$, *n* any integer, and for arbitrary $\mathcal{F}$ reduces to this case without much difficulty.

As a corollay to 1. above, if *f* is a projective morphism from a noetherian scheme to a noetherian ring, then the higher direct image $R^p f_* \mathcal{F}$ is coherent. The same result holds for proper morphisms *f*, as can be shown with the aid of Chow's lemma.

Sheaf cohomology groups H^i on a noetherian topological space vanish for *i* strictly greater than the dimension of the space. Thus the quantity, called the Euler characteristic of $\mathcal{F}$,

$$\chi(\mathcal{F}) = \sum_{i=0}^{\infty} (-1)^i \dim H^i(X, \mathcal{F})$$

is a well-defined integer (for *X* projective). One can then show $\chi(\mathcal{F}(n)) = P(n)$ for some polynomial *P* over rational numbers. Applying this procedure to the structure sheaf $\mathcal{O}_X$, one recovers the Hilbert polynomial of *X*. In particular, if *X* is irreducible and has dimension *r*, the arithmetic genus of *X* is given by

$$(-1)^r (\chi(\mathcal{O}_X) - 1),$$

which is manifestly intrinsic; i.e., independent of the embedding.

The arithmetic genus of a hypersurface of degree *d* is $\binom{d-1}{n}$ in $\mathbf{P}^n$. In particular, a smooth curve of degree *d* in P^2 has arithmetic genus $(d-1)(d-2)/2$. This is the genus formula.

Smooth Projective Varieties

Let *X* be a smooth projective variety where all of its irreducible components have dimension *n*. In this situation, the canonical sheaf ω_X, defined as the sheaf of Kähler differentials of top degree (i.e., algebraic *n*-forms), is a line bundle.

Serre Duality

Serre duality states that for any locally free sheaf $\mathcal{F}$ on *X*,

$$H^i(X, \mathcal{F}) \simeq H^{n-i}(X, \mathcal{F}^\vee \otimes \omega_X)'$$

where the superscript prime refers to the dual space and $\mathcal{F}^\vee$ is the dual sheaf of $\mathcal{F}$. A generalization to projective, but not necessarily smooth schemes is known as Verdier duality.

Riemann-Roch Theorem

For a (smooth projective) curve X, H^2 and higher vanish for dimensional reason and the space of the global sections of the structure sheaf is one-dimensional. Thus the arithmetic genus of X is the dimension of $H^1(X, \mathcal{O}_X)$. By definition, the geometric genus of X is the dimension of $H^0(X, \omega_X)$. Serre duality thus implies that the arithmetic genus and the geometric genus coincide. They will simply be called the genus of X.

Serre duality is also a key ingredient in the proof of the Riemann–Roch theorem. Since X is smooth, there is an isomorphism of groups

$$\mathrm{Cl}(X) \to \mathrm{Pic}(X), D \mapsto \mathcal{O}(D)$$

from the group of (Weil) divisors modulo principal divisors to the group of isomorphism classes of line bundles. A divisor corresponding to ω_X is called the canonical divisor and is denoted by K. Let $l(D)$ be the dimension of $H^0(X, \mathcal{O}(D))$. Then the Riemann–Roch theorem states: if g is a genus of X,

$$l(D) - l(K - D) = \deg D + 1 - g$$

for any divisor D on X. By the Serre duality, this is the same as:

$$\chi(\mathcal{O}(D)) = \deg D + 1 - g \,,$$

which can be readily proved. A generalization of the Riemann-Roch theorem to higher dimension is the Hirzebruch-Riemann-Roch theorem, as well as the far-reaching Grothendieck-Riemann-Roch theorem.

Hilbert Schemes

Hilbert schemes parametrize all closed subvarieties of a projective scheme X in the sense that the points (in the functorial sense) of H correspond to the closed subschemes of X. As such, the Hilbert scheme is an example of a moduli space, i.e., a geometric object whose points parametrize other geometric objects. More precisely, the Hilbert scheme parametrizes closed subvarieties whose Hilbert polynomial equals a prescribed polynomial P. It is a deep theorem of Grothendieck that there is a scheme H_X^P over k such that, for any k-scheme T, there is a bijection

$$\{morphisms\ T \to H_X^P\} \longleftrightarrow \{closed\ subschemes\ of\ X \times_k T\ flat\ over\ T,\ with\ Hilbert\ polynomial\ P.\}$$

The closed subscheme of $X \times H_X^P$ that corresponds to the identity map $H_X^P \to H_X^P$ is called the *universal family*.

For $P(z) = \binom{z+r}{r}$, the Hilbert scheme $H_{\mathbf{P}^n}^P$ is called the Grassmannian of r-planes in $\mathbf{P}^n$ and, if X is a projective scheme, H_X^P is called the Fano scheme of r-planes on X.

Complex Projective Varieties

In this section, all algebraic varieties are complex algebraic varieties. A key feature of the theory of complex projective varieties is the combination of algebraic and analytic methods. The transition between these theories is provided by the following link: since any complex polynomial is also a holomorphic function, any complex variety X yields a complex analytic space, denoted $X(\mathbf{C})$. Moreover, geometric properties of X are reflected by the ones of $X(\mathbf{C})$. For example, the latter is a complex manifold iff X is smooth; it is compact iff X is proper over C.

Relation to Complex Kähler Manifolds

Complex projective space is a Kähler manifold. This implies that, for any projective algebraic variety X, X(C) is a compact Kähler manifold. The converse is not in general true, but the Kodaira embedding theorem gives a criterion for a Kähler manifold to be projective.

In low dimensions, there are the following results:

- (Riemann) A compact Riemann surface (i.e., compact complex manifold of dimension one) is a projective variety. By the Torelli theorem, it is uniquely determined by its Jacobian.
- (Chow-Kodaira) A compact complex manifold of dimension two with two algebraically independent meromorphic functions is a projective variety.

GAGA and Chow's Theorem

Chow's theorem provides a striking way to go the other way, from analytic to algebraic geometry. It states that every analytic subvariety of a complex projective space is algebraic. The theorem may be interpreted to saying that a holomorphic function satisfying certain growth condition is necessarily algebraic: "projective" provides this growth condition. One can deduce from the theorem the following:

- Meromorphic functions on the complex projective space are rational.
- If an algebraic map between algebraic varieties is an analytic isomorphism, then it is an (algebraic) isomorphism. (This part is a basic fact in complex analysis.) In particular, Chow's theorem implies that a holomorphic map between projective varieties is algebraic. (consider the graph of such a map.)
- Every holomorphic vector bundle on a projective variety is induced by a unique algebraic vector bundle.
- Every holomorphic line bundle on a projective variety is a line bundle of a divisor.

Chow's theorem can be shown via Serre's GAGA principle. Its main theorem states:

> Let X be a projective scheme over C. Then the functor associating the coherent sheaves on X to the coherent sheaves on the corresponding complex analytic space X^{an} is an equivalence of categories. Furthermore, the natural maps
>
> $$H^i(X,\mathcal{F}) \to H^i(X^{an},\mathcal{F})$$

are isomorphisms for all i and all coherent sheaves $\mathcal{F}$ on X.

Complex Tori vs. Complex Abelian Varieties

The complex manifold associated to an abelian variety *A* over C is a compact complex Lie group. These can be shown to be of the form

$$\mathbb{C}^g / L$$

and are also referred to as complex tori. Here, *g* is the dimension of the torus and *L* is a lattice (also referred to as period lattice).

According to the uniformization theorem already mentioned above, any torus of dimension 1 arises from an abelian variety of dimension 1, i.e., from an elliptic curve. In fact, the Weierstrass's elliptic function $\wp$ attached to *L* satisfies a certain differential equation and as a consequence it defines a closed immersion:

$$\mathbb{C} / L \to \mathbf{P}^2, L \mapsto (0:0:1), z \mapsto (1:\wp(z):\wp'(z)).$$

There is a *p*-adic analog, the p-adic uniformization theorem.

For higher dimensions, the notions of complex abelian varieties and complex tori differ: only polarized complex tori come from abelian varieties.

Kodaira Vanishing

The fundamental Kodaira vanishing theorem states that for an ample line bundle $\mathcal{L}$ on a smooth projective variety *X* over a field of characteristic zero,

$$H^i(X, \mathcal{L} \otimes \omega_X) = 0$$

for $i > 0$, or, equivalently by Serre duality $H^i(X, \mathcal{L}^{-1}) = 0$ for $i < n$. The first proof of this theorem used analytic methods of Kähler geometry, but a purely algebraic proof was found later. The Kodaira vanishing in general fails for a smooth projective variety in positive characteristic. Kodaira's theorem is one of various vanishing theorems, which give criteria for higher sheaf cohomologies to vanish. Since the Euler characteristic of a sheaf is often more manageable than individual cohomology groups, this often has important consequences about the geometry of projective varieties.

Related Notions

- Multi-projective variety
- *Weighted projective variety*, a closed subvariety of a weighted projective space

Quasi-projective Variety

In mathematics, a quasi-projective variety in algebraic geometry is a locally closed subset of a projective variety, i.e., the intersection inside some projective space of a Zariski-open and a Zariski-closed subset. A similar definition is used in scheme theory, where a *quasi-projective scheme* is a locally closed subscheme of some projective space.

Relationship to Affine Varieties

An affine space is a Zariski-open subset of a projective space, and since any closed affine subset U can be expressed as an intersection of the projective completion $\bar{U}$ and the affine space embedded in the projective space, this implies that any affine variety is quasiprojective. There are locally closed subsets of projective space that are not affine, so that quasi-projective is more general than affine. Taking the complement of a single point in projective space of dimension at least 2 gives a non-affine quasi-projective variety. This is also an example of a quasi-projective variety that is neither affine nor projective.

Examples

Since quasi-projective varieties generalize both affine and projective varieties, they are sometimes referred to simply as *varieties*. Varieties isomorphic to affine algebraic varieties as quasi-projective varieties are called affine varieties; similarly for projective varictics. For example, the complement of a point in the affine line, i.e., $X = \mathbb{A}^1 \setminus \{0\}$, is isomorphic to the zero set of the polynomial $xy-1$ in the affine plane. As an affine set X is not closed since any polynomial zero on the complement must be zero on the affine line. For another example, the complement of any conic in projective space of dimension 2 is affine. Varieties isomorphic to open subsets of affine varieties are called quasi-affine.

Quasi-projective varieties are *locally affine* in the sense that a manifold is locally Euclidean — every point of a quasiprojective variety has a neighborhood given by an affine variety. This yields a basis of affine sets for the Zariski topology on a quasi-projective variety.

Hilbert's Nullstellensatz

Hilbert's Nullstellensatz (German for "theorem of zeros," or more literally, "zero-locus-theorem") is a theorem that establishes a fundamental relationship between geometry and algebra. This relationship is the basis of algebraic geometry, a branch of mathematics. It relates algebraic sets to ideals in polynomial rings over algebraically closed fields. This relationship was discovered by David Hilbert who proved the Nullstellensatz and several other important related theorems named after him (like Hilbert's basis theorem).

Formulation

Let k be a field (such as the rational numbers) and K be an algebraically closed field extension (such as the complex numbers), consider the polynomial ring $k[X_1,X_2,\ldots, X_n]$ and let I be an ideal in this ring. The algebraic set V(I) defined by this ideal consists of all n-tuples x = $(x_1,\ldots,x_n)$ in K^n such that f(x) = 0 for all f in I. Hilbert's Nullstellensatz states that if p is some polynomial in $k[X_1,X_2,\ldots, X_n]$ that vanishes on the algebraic set V(I), i.e. p(x) = 0 for all x in $V(I)$, then there exists a natural number r such that p^r is in I.

An immediate corollary is the "weak Nullstellensatz": The ideal I in $k[X_1,X_2,\ldots, X_n]$ contains 1 if and only if the polynomials in I do not have any common zeros in K^n. It may also be formulated as follows: if I is a proper ideal in $k[X_1,X_2,..., X_n]$, then V(I) cannot be empty, i.e. there exists a common zero for all the polynomials in the ideal in every algebraically closed extension of k. This is the reason for the name of the theorem, which can be proved easily from the 'weak' form using the Rabinowitsch trick. The assumption of considering common zeros in an algebraically closed field is essential here; for example, the elements of the proper ideal $(X^2 + 1)$ in R[X] do not have a common zero in R. With the notation common in algebraic geometry, the Nullstellensatz can also be formulated as

$$I(V(J)) = \sqrt{J}$$

for every ideal J. Here, $\sqrt{J}$ denotes the radical of J and I(U) is the ideal of all polynomials that vanish on the set U.

In this way, we obtain an order-reversing bijective correspondence between the algebraic sets in K^n and the radical ideals of $K[X_1,X_2,..., X_n]$. In fact, more generally, one has a Galois connection between subsets of the space and subsets of the algebra, where "Zariski closure" and "radical of the ideal generated" are the closure operators.

As a particular example, consider a point $P = (a_1,\ldots,a_n) \in K^n$. Then $I(P) = (X_1 - a_1,\ldots,X_n - a_n)$. More generally,

$$\sqrt{I} = \bigcap_{(a_1,\ldots,a_n)\in V(I)} (X_1 - a_1,\ldots,X_n - a_n).$$

Conversely, every maximal ideal of the polynomial ring $K[X_1,\ldots,X_n]$ (note that K is algebraically closed) is of the form $(X_1 - a_1,\ldots,X_n - a_n)$ for some $a_1,\ldots,a_n \in K$.

As another example, an algebraic subset W in K^n is irreducible (in the Zariski topology) if and only if $I(W)$ is a prime ideal.

Proof and Generalization

There are many known proofs of the theorem. One proof uses Zariski's lemma, which asserts that an associative algebra that is finitely generated over a field k is a field is a finite field extension of k.

Let $A = k[t_1,\ldots,t_n]$ (k algebraically closed field), I an ideal of A and V the common zeros of I in k^n. Clearly, $\sqrt{I} \subseteq I(V)$. Let $f \notin \sqrt{I}$. Then $f \notin \mathfrak{p}$ for some prime ideal $\mathfrak{p} \supseteq I$ in A. Let $R = (A/\mathfrak{p})[f^{-1}]$ and $\mathfrak{m}$ a maximal ideal in R. By Zariski's lemma, $R/\mathfrak{m}$ is a finite extension of k; thus, is k since k is algebraically closed. Let x_i be the images of t_i under the natural map $A \to k$. It follows that $x = (x_1,\ldots,x_n) \in V$ and $f(x) \neq 0$.

The Nullstellensatz will also follow trivially once one systematically developed the theory of a Jacobson ring, a ring in which a radical ideal is an intersection of maximal ideals. Let R be a Jacobson ring. If S is a finitely generated R-algebra, then S is a Jacobson ring. Further, if $\mathfrak{n} \subset S$ is a maximal ideal, then $\mathfrak{m} := \mathfrak{n} \cap R$ is a maximal ideal of R, and $S/\mathfrak{n}$ is a finite extension field of $R/\mathfrak{m}$.

Another generalization states that a faithfully flat morphism of schemes $f: Y \to X$ locally of finite type with X quasi-compact has a *quasi-section*, i.e. there exists X' affine and faithfully flat and quasi-finite over X together with an X-morphism $X' \to Y$

Effective Nullstellensatz

In all of its variants, Hilbert's Nullstellensatz asserts that some polynomial g belongs or not to an ideal generated, say, by $f_1, \ldots, f_k$; we have $g = f^r$ in the strong version, $g = 1$ in the weak form. This means the existence or the non-existence of polynomials $g_1, \ldots, g_k$ such that $g = f_1g_1 + \ldots + f_kg_k$. The usual proofs of the Nullstellensatz are not constructive, non-effective, in the sense that they do not give any way to compute the g_i.

It is thus a rather natural question to ask if there is an effective way to compute the g_i (and the exponent r in the strong form) or to prove that they do not exist. To solve this problem, it suffices to provide an upper bound on the total degree of the g_i: such a bound reduces the problem to a finite system of linear equations that may be solved by usual linear algebra techniques. Any such upper bound is called an effective Nullstellensatz.

A related problem is the ideal membership problem, which consists in testing if a polynomial belongs to an ideal. For this problem also, a solution is provided by an upper bound on the degree of the g_i. A general solution of the ideal membership problem provides an effective Nullstellensatz, at least for the weak form.

In 1925, Grete Hermann gave an upper bound for ideal membership problem that is doubly exponential in the number of variables. In 1982 Mayr and Meyer gave an example where the g_i have a degree that is at least double exponential, showing that every general upper bound for the ideal membership problem is doubly exponential in the number of variables.

Since most mathematicians at the time assumed the effective Nullstellensatz was at least as hard as ideal membership, few mathematicians sought a bound better than double exponential. In 1987, however, W. Dale Brownawell gave an upper bound for the effective Nullstellensatz that is simply exponential in the number of variables. Brownawell's proof relied on analytic techniques valid only in characteristic 0, but, one year later, János Kollár gave a purely algebraic proof, valid in any characteristic, of a slightly better bound.

In the case of the weak Nullstellensatz, Kollár's bound is the following:

> Let $f_1, \ldots, f_s$ be polynomials in $n \geq 2$ variables, of total degree $d_1 \geq \ldots \geq d_s$. If there exist polynomials g_i such that $f_1g_1 + \ldots + f_sg_s = 1$, then they can be chosen such that
>
> $$\deg(f_i g_i) \leq \max(d_s, 3) \prod_{j=1}^{\min(n,s)-1} \max(d_j, 3).$$
>
> This bound is optimal if all the degrees are greater than 2.

If d is the maximum of the degrees of the f_i, this bound may be simplified to

$$\max(3, d)^{\min(n,s)}.$$

Kollár's result has been improved by several authors. As of 14 October 2012, the best improvement, due to M. Sombra is

$$\deg(f_i g_i) \leq 2d_s \prod_{j=1}^{\min(n,s)-1} d_j.$$

His bound improves Kollár's as soon as at least two of the degrees that are involved are lower than 3.

Projective Nullstellensatz

We can formulate a certain correspondence between homogeneous ideals of polynomials and algebraic subsets of a projective space, called the projective Nullstellensatz, that is analogous to the affine one. To do that, we introduce some notations. Let $R = k[t_0,\ldots,t_n]$. The homogeneous ideal $R_+ = \bigoplus_{d\geq 1} R_d$ is called the *maximal homogeneous ideal*. As in the affine case, we let: for a subset $S \subseteq \mathbb{P}^n$ and a homogeneous ideal *I* of *R*,

$$\mathrm{I}_{\mathbb{P}^n}(S) = \{f \in R_+ \mid f = 0 \text{ on } S\},$$
$$\mathrm{V}_{\mathbb{P}^n}(I) = \{x \in \mathbb{P}^n \mid f(x) = 0 \text{ for all } f \in I\}.$$

By $f = 0 \text{ on } S$ we mean: for every homogeneous coordinates $(a_0 : \cdots : a_n)$ of a point of *S* we have $f(a_0,\ldots,a_n) = 0$. This implies that the homogeneous components of *f* are also zero on *S* and thus that $\mathrm{I}_{\mathbb{P}^n}(S)$ is a homogeneous ideal. Equivalently, $\mathrm{I}_{\mathbb{P}^n}(S)$ is the homogeneous ideal generated by homogeneous polynomials *f* that vanish on *S*. Now, for any homogeneous ideal $I \subseteq R_+$, by the usual Nullstellensatz, we have:

$$\sqrt{I} = \mathrm{I}_{\mathbb{P}^n}(\mathrm{V}_{\mathbb{P}^n}(I)),$$

and so, like in the affine case, we have:

> There exists an order-reversing one-to-one correspondence between proper homogeneous radical ideals of *R* and subsets of $\mathbb{P}^n$ of the form $\mathrm{V}_{\mathbb{P}^n}(I)$. The correspondence is given by $\mathrm{I}_{\mathbb{P}^n}$ and $\mathrm{V}_{\mathbb{P}^n}$.

Analytic Nullstellensatz

The Nullstellensatz also holds for the germs of holomorphic functions at a point of complex *n*-space $\mathbb{C}^n$. Precisely, for each open subset $U \subset \mathbb{C}^n$, let $\mathcal{O}_{\mathbb{C}^n}(U)$ denote the ring of holomorphic functions on *U*; then $\mathcal{O}_{\mathbb{C}^n}$ is a *sheaf* on $\mathbb{C}^n$. The stalk $\mathcal{O}_{\mathbb{C}^n,0}$ at, say, the origin can be shown to be a Noetherian local ring that is a unique factorization domain.

If $f \in \mathcal{O}_{\mathbb{C}^n,0}$ is a germ represented by a holomorphic function $\tilde{f} : U \to \mathbb{C}$, then let

$V_0(f)$ = the equivalence class of the set $\{z \in U \mid \tilde{f}(z) = 0\}$

where two subsets *X*, *Y* of $\mathbb{C}^n$ are considered equivalent if $X \cap U = Y \cap U$ for some neighborhood *U* of 0. Note $V_0(f)$ is independent of a choice of the representative $\tilde{f}$. For each ideal $I \subset \mathcal{O}_{\mathbb{C}^n,0}$, let $V_0(I)$ denote $V_0(f_1) \cap \ldots \cap V_0(f_r)$ for some generators $f_1,\ldots,f_r$ of *I*. It is well-defined; i.e., is independent of a choice of the generators.

For each subset $X \subset \mathbb{C}^n$, let $I_0(X) = \{f \in \mathcal{O}_{\mathbb{C}^n,0} \mid V_0(f) \supset X\}$. It is easy to see that $I_0(X)$ is an ideal of $\mathcal{O}_{\mathbb{C}^n,0}$ and that $I_0(X) = I_0(Y)$ if $X \sim Y$ in the sense discussed above.

The analytic Nullstellensatz then states: for each ideal $I \subset \mathcal{O}_{\mathbb{C}^n,0}$,

$$\sqrt{I} = I_0(V_0(I))$$

where the left-hand side is the radical of *I*.

Zariski's Main Theorem

In algebraic geometry, Zariski's main theorem, proved by Oscar Zariski (1943), is a statement about the structure of birational morphisms stating roughly that there is only one branch at any normal point of a variety. It is the special case of Zariski's connectedness theorem when the two varieties are birational.

Zariski's main theorem can be stated in several ways which at first sight seem to be quite different, but are in fact deeply related. Some of the variations that have been called Zariski's main theorem are as follows:

- The total transform of a normal fundamental point of a birational map has positive dimension. This is essentially Zariski's original form of his main theorem.
- A birational morphism with finite fibers to a normal variety is an isomorphism to an open subset.
- The total transform of a normal point under a proper birational morphism is connected.
- A closely related theorem of Grothendieck describes the structure of quasi-finite morphisms of schemes, which implies Zariski's original main theorem.
- Several results in commutative algebra that imply the geometric form of Zariski's main theorem.
- A normal local ring is unibranch, which is a variation of the statement that the transform of a normal point is connected.
- The local ring of a normal point of a variety is analytically normal. This is a strong form of the statement that it is unibranch.

The name "Zariski's main theorem" comes from the fact that Zariski labelled it as the "MAIN THEOREM" in Zariski (1943).

Zariski's Main Theorem for Birational Morphisms

Let *f* be a birational mapping of algebraic varieties *V* and *W*. Recall that *f* is defined by a closed subvariety $\Gamma \subset V \times W$ (a "graph" of *f*) such that the projection on the first factor p_1 induces an isomorphism between an open $U \subset V$ and $p_1^{-1}(U)$, and such that $p_2 \circ p_1^{-1}$ is an isomorphism on *U* too.

The complement of U in V is called a *fundamental variety* or *indeteminancy locus*, and an image of a subset of V under $p_2 \circ p_1$ is called a *total transform* of it.

The original statement of the theorem reads:

> MAIN THEOREM: If W is an irreducible fundamental variety on V of a birational correspondence T between V and V' and if T has no fundamental elements on V' then — under the assumption that V is locally normal at W — each irreducible component of the transform $T[W]$ is of higher dimension than W.

Here T is essentially a morphism from V' to V that is birational, W is a subvariety of the set where the inverse of T is not defined whose local ring is normal, and the transform $T[W]$ means the inverse image of W under the morphism from V' to V.

Here are some variants of this theorem stated using more recent terminology. Hartshorne calls the following connectedness statement "Zariski's Main theorem":

> If $f{:}X \to Y$ is a birational projective morphism between noetherian integral schemes, then the inverse image of every normal point of Y is connected.

The following consequence of it also goes under this name:

> If $f{:}X \to Y$ is a birational transformation of projective varieties with Y normal, then the total transform of a fundamental point of f is connected and of dimension at least 1.

Examples

- Suppose that V is a smooth variety of dimension greater than 1 and V' is given by blowing up a point W on V. Then V is normal at W, and the component of the transform of W is a projective space, which has dimension greater than W as predicted by Zariski's original form of his main theorem.
- In the previous example the transform of W was irreducible. It is easy to find examples where the total transform is reducible by blowing up other points on the transform. For example, if V' is given by blowing up a point W on V and then blowing up another point on this transform, the total transform of W has 2 irreducible components meeting at a point. As predicted by Hartshorne's form of the main theorem, the total transform is connected and of dimension at least 1.
- For an example where W is not normal and the conclusion of the main theorem fails, take V' to be a smooth variety, and take V to be given by identifying two distinct points on V', and take W to be the image of these two points. Then W is not normal, and the transform of W consists of two points, which is not connected and does not have positive dimension.

Zariski's Main Theorem for Quasifinite Morphisms

In EGA III, Grothendieck calls the following statement which does not involve connectedness a "Main theorem" of Zariski Grothendieck:

> If $f:X \to Y$ is a quasi-projective morphism of Noetherian schemes then the set of points that are isolated in their fiber is open in X. Moreover the induced scheme of this set is isomorphic to an open subset of a scheme that is finite over Y.

In EGA IV, Grothendieck observed that the last statement could be deduced from a more general theorem about the structure of quasi-finite morphisms, and the latter is often referred to as the "Zariski's main theorem in the form of Grothendieck". It is well known that open immersions and finite morphisms are quasi-finite. Grothendieck proved that under the hypothesis of separatedness all quasi-finite morphisms are compositions of such Grothendieck:

> if Y is a quasi-compact separated scheme and $f: X \to Y$ is a separated, quasi-finite, finitely presented morphism then there is a factorization into $X \to Z \to Y$, where the first map is an open immersion and the second one is finite.

The relation between this theorem about quasi-finite morphisms and Théorème 4.4.3 of EGA III quoted above is that if $f:X \to Y$ is a projective morphism of varieties, then the set of points that are isolated in their fiber is quasifinite over Y. Then structure theorem for quasi-finite morphisms applies and yields the desired result.

Zariski's Main Theorem for Commutative Rings

Zariski (1949) reformulated his main theorem in terms of commutative algebra as a statement about local rings. Grothendieck generalized Zariski's formulation as follows:

> If B is an algebra of finite type over a local Noetherian ring A, and n is a maximal ideal of B which is minimal among ideals of B whose inverse image in A is the maximal ideal m of A, then there is a finite A-algebra A' with a maximal ideal m' (whose inverse image in A is m) such that the localization B_n is isomorphic to the A-algebra $A'_{m'}$.

If in addition A and B are integral and have the same field of fractions, and A is integrally closed, then this theorem implies that A and B are equal. This is essentially Zariski's formulation of his main theorem in terms of commutative rings.

Zariski's Main Theorem: Topological Form

A topological version of Zariski's main theorem says that if x is a (closed) point of a normal complex variety it is unibranch; in other words there are arbitrarily small neighborhoods U of x such that the set of non-singular points of U is connected.

The property of being normal is stronger than the property of being unibranch: for example, a cusp of a plane curve is unibranch but not normal.

Zariski's Main Theorem: Power Series Form

A formal power series version of Zariski's main theorem says that if x is a normal point of a variety then it is analytically normal; in other words the completion of the local ring at x is a normal integral domain.

AF+BG Theorem

In algebraic geometry, a field of mathematics, the AF+BG theorem (also known as Max Noether's fundamental theorem) is a result of Max Noether which describes when the equation of an algebraic curve in the complex projective plane can be written in terms of the equations of two other algebraic curves.

Statement

Let F, G, and H be homogeneous polynomials in three variables, with H having higher degree than F and G; let $a = \deg H - \deg F$ and $b = \deg H - \deg G$ (both positive integers) be the differences of the degrees of the polynomials. Suppose that the greatest common divisor of F and G is constant, which means that the projective curves that they define in the projective plane P^2 have an intersection consisting in a finite number of points. For each point P of this intersection, the polynomials F and G generate an ideal $(F, G)_P$ of the local ring of P^2 at P (this local ring is the ring of the fractions n/d, where n and d are polynomials in three variables and $d(P) \neq 0$). The theorem asserts that, if H lies in $(F, G)_P$ for every intersection point P, then there are homogeneous polynomials A and B of degrees a and b, respectively, such that $H = AF + BG$. Furthermore, any two choices of A differ by a multiple of G, and similarly any two choices of B differ by a multiple of F.

Related Results

This theorem may be viewed as a generalization of Bézout's identity, which provides a condition under which an integer or a univariate polynomial h may be expressed as an element of the ideal generated by two other integers or univariate polynomials f and g: such a representation exists exactly when h is a multiple of the greatest common divisor of f and g. The AF+BG condition expresses, in terms of divisors (sets of points, with multiplicities), a similar condition under which a homogeneous polynomial H in three variables can be written as an element of the ideal generated by two other polynomials F and G.

This theorem is also a refinement, for this particular case, of Hilbert's Nullstellensatz, which provides a condition expressing that some power of a polynomial h (in any number of variables) belongs to the ideal generated by a finite set of polynomials.

Grothendieck Trace Formula

In algebraic geometry, the Grothendieck trace formula expresses the number of points of a variety over a finite field in terms of the trace of the Frobenius endomorphism on its cohomology groups. There are several generalizations: the Frobenius endomorphism can be replaced by a more general endomorphism, in which case the points over a finite field are replaced by its fixed points, and there is also a more general version for a sheaf over the variety, where the cohomology groups are replaced by cohomology with coefficients in the sheaf.

The Grothendieck trace formula is an analogue in algebraic geometry of the Lefschetz fixed-point theorem in algebraic topology.

One application of the Grothendieck trace formula is to express the zeta function of a variety over a finite field, or more generally the L-series of a sheaf, as a sum over traces of Frobenius on cohomology groups. This is one of the steps used in the proof of the Weil conjectures.

Behrend's trace formula generalizes the formula to algebraic stacks.

Formal Statement for L-functions

Let k be a finite field, l a prime number invertible in k, X a smooth k-scheme of dimension n, and $\mathcal{F}$ a constructible $\mathbb{Q}_l$-sheaf on X. Then the following cohomological expression for the L-function of $\mathcal{F}$ holds:

$$L(X,\mathcal{F},t)=\prod_{i=0}^{2n}\det(1-t\cdot F\mid H_c^i(X_{\bar{k}},\mathcal{F}))^{(-1)^{i+1}}=\frac{\det(1-t\cdot F\mid H_c^1(X_{\bar{k}},\mathcal{F}))\ldots\det(1-t\cdot F\mid H_c^{2n-1}(X_{\bar{k}},\mathcal{F}))}{\det(1-t\cdot F\mid H_c^0(X_{\bar{k}},\mathcal{F}))\ldots\det(1-t\cdot F\mid H_c^{2n}(X_{\bar{k}},\mathcal{F}))}$$

where F is everywhere a geometric Frobenius action on l-adic cohomology with compact supports of the sheaf $\mathcal{F}$. Taking logarithmic derivatives of both formal power series produces a statement on sums of traces for each finite field extension E of the base field k:

$$\sum_{x\in X(E)}\operatorname{tr}(F_E\mid\mathcal{F}_x)=\sum_{i=0}^{2n}\operatorname{tr}(F_E\mid H_c^i(X_{\bar{k}},\mathcal{F}))$$

For a constant sheaf $\mathbb{Q}_l$ (viewed as $(\varprojlim\mathbb{Z}/l^n\mathbb{Z})\otimes\mathbb{Q}$ to qualify as an l-adic sheaf) the left hand side of this formula is the number of E-points of X.

Behrend's Trace Formula

In algebraic geometry, Behrend's formula is a generalization of the Grothendieck–Lefschetz trace formula to a smooth algebraic stack over a finite field, conjectured in 1993 and proven in 2003 by Kai Behrend. Unlike the classical one, the formula counts points in the "stacky way"; it takes into account the presence of nontrivial automorphisms.

The desire for the formula comes from the fact that it applies to the moduli stack of principal bundles on a curve over a finite field (in some instances indirectly, via the Harder–Narasimhan stratification, as the moduli stack is not of finite type.).

Deligne found an example that shows the formula may be interpreted as a sort of the Selberg trace formula.

A proof of the formula in the context of the six operations formalism developed by Laszlo and Olsson is given by Shenghao Sun.

Formulation

By definition, if C is a category in which each object has finitely many automorphisms, the number of points in C is denoted by

$$\#C = \sum_p \frac{1}{\#\mathrm{Aut}(p)},$$

with the sum running over representatives p of all isomorphism classes in C. (The series may diverge in general.) The formula states: for a smooth algebraic stack X of finite type over a finite field $\mathbf{F}_q$ and the "arithmetic" Frobenius $\phi^{-1}: X \to X$, i.e., the inverse of the usual geometric Frobenius ϕ in Grothendieck's formula,

$$\# X(\mathbf{F}_q) = q^{\dim X} \sum_{i=0}^{\infty} (-1)^i \operatorname{tr}(\phi^{-1}; H^i(X, \mathbb{Q}_l)).$$

Here, it is crucial that the cohomology of a stack is with respect to the smooth topology (not etale).

When X is a variety, the smooth cohomology is the same as etale one and, via the Poincaré duality, this is equivalent to Grothendieck's trace formula. (But the proof relies on Grothendieck's formula, so this does not subsume Grothendieck's.)

Simple Example

Consider $B\mathbb{G}_m = [\operatorname{Spec} \mathbf{F}_q / \mathbb{G}_m]$, the classifying stack of the multiplicative group scheme (that is, $\mathbb{G}_m(R) = R^\times$). By definition, $B\mathbb{G}_m(\mathbf{F}_q)$ is the category of principal $\mathbb{G}_m$-bundles over $\operatorname{Spec} \mathbf{F}_q$, which has only one isomorphism class (since all such bundles are trivial by Lang's theorem). Its group of automorphisms is $\mathbb{G}_m$, which means that the number of $\mathbf{F}_q$-isomorphisms is $\#\mathbb{G}_m(\mathbf{F}_q) = \#\mathbf{F}_q^\times = q - 1$.

On the other hand, we may compute the l-adic cohomology of $B\mathbb{G}_m$ directly. We remark that in the topological setting, we have $B\mathbb{C}^\times \cong \mathbb{CP}^\infty$ (where $B\mathbb{C}^\times$ now denotes the usual classifying space of a topological group), whose rational cohomology ring is a polynomial ring in one generator (Borel's theorem), but we shall not use this directly. If we wish to stay in the world of algebraic geometry, we may instead "approximate" $B\mathbb{G}_m$ by projective spaces of larger and larger dimension. Thus we consider the map $B\mathbb{G}_m \to \mathbb{P}^N$ induced by the $\mathbb{G}_m$-bundle corresponding to $\mathcal{O}(1)$. This map induces an isomorphism in cohomology in degrees up to $2N$. Thus the even (resp. odd) Betti numbers of $B\mathbb{G}_m$ are 1 (resp. 0), and the l-adic Galois representation on the $(2n)$th cohomology group is the nth power of the cyclotomic character. The second part is a consequence of the fact that the cohomology of $\mathbb{P}^N$ is generated by algebraic cycle classes. This shows that

$$\sum_{i \ge 0} (-1)^i \operatorname{tr}(\phi^{-1}; H^i(B\mathbb{G}_m, \mathbb{Q}_l)) = 1 + 1/q + 1/q^2 + \cdots = \frac{q}{q-1}.$$

Note that $\dim B\mathbb{G}_m = \dim \operatorname{Spec} \mathbf{F}_q - \dim \mathbb{G}_m = -1$. Multiplying by q^{-1}, one obtains the predicted equality.

Bézout's Theorem

Bézout's theorem is a statement in algebraic geometry concerning the number of common points, or intersection points, of two plane algebraic curves which do not share a common component (that is, which do not have infinitely many common points). The theorem states that the number of common points of two such curves is at most equal to the product of their degrees, and equality holds if one counts points at infinity and points with complex coordinates (or more generally, coordinates from the algebraic closure of the ground field), and if each point is counted with its intersection multiplicity.

Bézout's theorem refers also to the generalization to higher dimensions: Let there be n homogeneous polynomials in $n+1$ variables, of degrees $d_1,\ldots,d_n$, that define n hypersurfaces in the projective space of dimension n. If the number of intersection points of the hypersurfaces is finite over an algebraic closure of the ground field, then this number is $d_1 \cdots d_n$, if the points are counted with their multiplicity. As in the case of two variables, in the case of affine hypersurfaces, and when not counting multiplicities nor non-real points, this theorem provides only an upper bound of the number of points, which is often reached. This is often referred to as Bézout's bound.

Bézout's theorem is fundamental in computer algebra and effective algebraic geometry, by showing that most problems have a computational complexity that is at least exponential in the number of variables. It follows that in these areas, the best complexity that may be hoped for will occur in algorithms that have a complexity which is polynomial in Bézout's bound.

Rigorous Statement

Suppose that X and Y are two plane projective curves defined over a field F that do not have a common component (this condition means that X and Y are defined by polynomials, whose polynomial greatest common divisor is a constant; in particular, it holds for a pair of "generic" curves). Then the total number of intersection points of X and Y with coordinates in an algebraically closed field E which contains F, counted with their multiplicities, is equal to the product of the degrees of X and Y.

The generalization in higher dimension may be stated as:

Let n projective hypersurfaces be given in a projective space of dimension n over an algebraic closed field, which are defined by n homogeneous polynomials in $n + 1$ variables, of degrees $d_1,\ldots,d_n$. Then either the number of intersection points is infinite, or the number of intersection points, counted with multiplicity, is equal to the product $d_1 \cdots d_n$. If the hypersurfaces are irreducible and in relative general position, then there are $d_1 \cdots d_n$ intersection points, all with multiplicity 1.

There are various proofs of this theorem. In particular, it may be deduced by applying iteratively the following generalization: if V is a projective algebraic set of dimension δ and degree d_1, and H is a hypersurface (defined by a polynomial) of degree d_2, that does not contain any irreducible component of V, then the intersection of V and H has dimension $\delta-1$ and degree $d_1 d_2$.

Bézout's theorem has been further generalized as the so-called multi-homogeneous Bézout theorem.

Intersection Multiplicity

The most delicate part of Bézout's theorem and its generalization to the case of k algebraic hypersurfaces in k-dimensional projective space is the procedure of assigning the proper intersection multiplicities. If P is a common point of two plane algebraic curves X and Y that is a non-singular point of both of them and, moreover, the tangent lines to X and Y at P are distinct then the intersection multiplicity is one. This corresponds to the case of "transversal intersection". If the curves X and Y have a common tangent at P then the multiplicity is at least two.

Examples

- Two distinct non-parallel lines (in the same plane) always meet in exactly one point. Two parallel lines intersect at a unique point that lies at infinity. To see how this works algebraically, in projective space, the lines $x+2y=3$ and $x+2y=5$ are represented by the homogeneous equations $x+2y-3z=0$ and $x+2y-5z=0$. Solving, we get $x=-2y$ and $z=0$, corresponding to the point (-2:1:0) in homogeneous coordinates. As the z-coordinate is 0, this point lies on the line at infinity.
- The special case where one of the curves is a line can be derived from the fundamental theorem of algebra. In this case the theorem states that an algebraic curve of degree n intersects a given line in n points, counting the multiplicities. For example, the parabola defined by $y-x^2=0$ has degree 2; the line $y-ax=0$ has degree 1, and they meet in exactly two points when $a \neq 0$ and touch at the origin (intersect with multiplicity two) when $a = 0$.
- Two conic sections generally intersect in four points, some of which may coincide. To properly account for all intersection points, it may be necessary to allow complex coordinates and include the points on the infinite line in the projective plane. For example:
    - Two circles never intersect in more than two points in the plane, while Bézout's theorem predicts four. The discrepancy comes from the fact that every circle passes through the same two complex points on the line at infinity. Writing the circle

$$(x-a)^2+(y-b)^2=r^2$$

in homogeneous coordinates, we get

$$(x-az)^2+(y-bz)^2-r^2z^2=0$$

from which it is clear that the two points $(1:i:0)$ and $(1:-i:0)$ lie on every circle. When two circles don't meet at all in the real plane, the two other intersections have non-zero imaginary parts, or if they are concentric then they meet at exactly the two points on the line at infinity with an intersection multiplicity of two.

- Any conic should meet the line at infinity at two points according to the theorem. A hyperbola meets it at two real points corresponding to the two directions of the asymptotes. An ellipse meets it at two complex points which are conjugate to one another---in the case of a circle, the points $(1:i:0)$ and $(1:-i:0)$. A parabola meets it at only one point, but it is a point of tangency and therefore counts twice.
    - The following pictures show examples in which the circle $x^2+y^2-1=0$ meets another ellipse in fewer intersection points because at least one of them has multiplicity greater than 1:

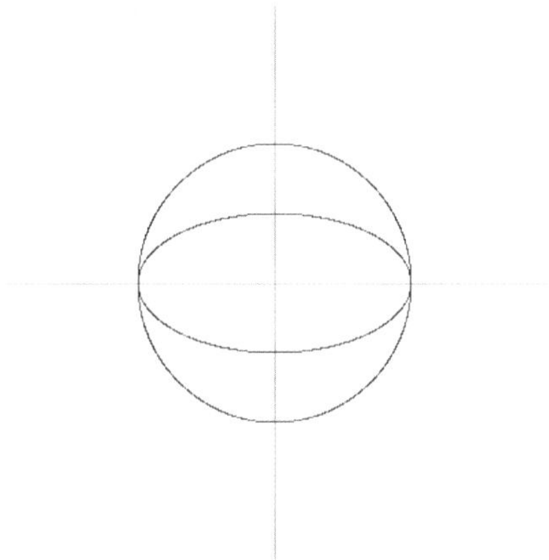

$x^2 + 4y^2 - 1 = 0$: two intersections of multiplicity 2

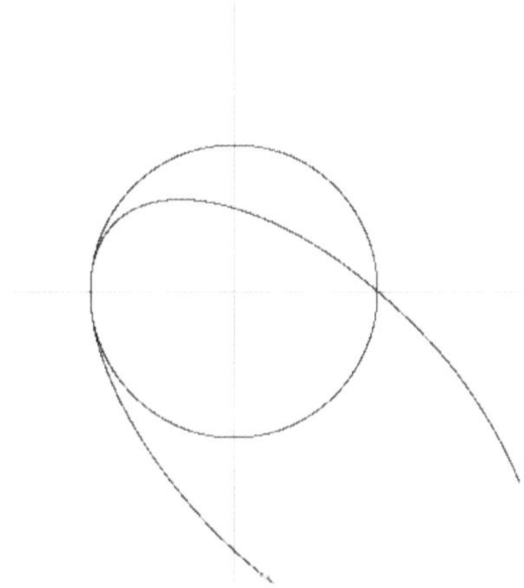

$5x^2 + 6xy + 5y^2 + 6y - 5 = 0$: an intersection of multiplicity 3

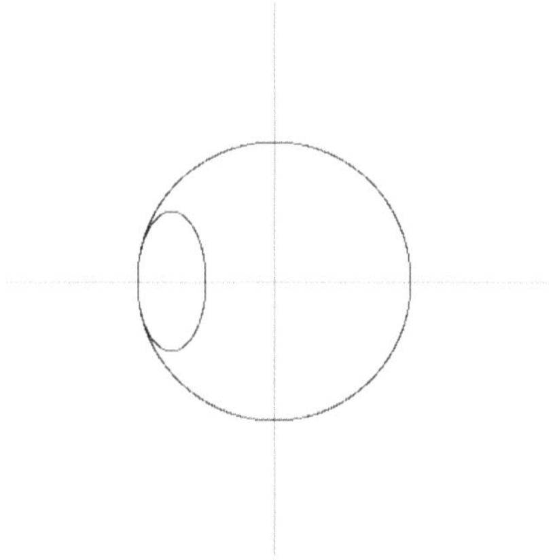

$4x^2 + y^2 + 6x + 2 = 0$: an intersection of multiplicity 4

Sketch of Proof

Write the equations for X and Y in homogeneous coordinates as

$$a_0 z^m + a_1 z^{m-1} + \ldots + a_{m-1} z + a_m = 0$$

$$b_0 z^n + b_1 z^{n-1} + \cdots + b_{n-1} z + b_n = 0$$

where a_i and b_i are homogeneous polynomials of degree i in x and y. The points of intersection of X and Y correspond to the solutions of the system of equations. Form the Sylvester matrix; in the case $m=4$, $n=3$ this is

$$S = \begin{pmatrix} a_0 & a_1 & a_2 & a_3 & a_4 & 0 & 0 \\ 0 & a_0 & a_1 & a_2 & a_3 & a_4 & 0 \\ 0 & 0 & a_0 & a_1 & a_2 & a_3 & a_4 \\ b_0 & b_1 & b_2 & b_3 & 0 & 0 & 0 \\ 0 & b_0 & b_1 & b_2 & b_3 & 0 & 0 \\ 0 & 0 & b_0 & b_1 & b_2 & b_3 & 0 \\ 0 & 0 & 0 & b_0 & b_1 & b_2 & b_3 \end{pmatrix}.$$

The determinant $|S|$ of S, which is also called the resultant of the two polynomials, is 0 exactly when the two equations have a common solution in z. The terms of $|S|$, for example $(a_0)^n(b_n)^m$, all have degree mn, so $|S|$ is a homogeneous polynomial of degree mn in x and y (recall that a_i and b_i are themselves polynomials). By the fundamental theorem of algebra, this can be factored into mn linear factors so there are mn solutions to the system of equations. The linear factors correspond to the lines that join the origin to the points of intersection of the curves.

References

- Dieudonné, Jean (1972). "The historical development of algebraic geometry". The American Mathematical Monthly. 79 (8): 827–866. doi:10.2307/2317664. JSTOR 2317664
- Drton, Mathias; Sturmfels, Bernd; Sullivant, Seth (2009). Lectures on Algebraic Statistics. Springer. ISBN 978-3-7643-8904-8
- Dolgachev, Igor (1982), "Weighted projective varieties", Group actions and vector fields (Vancouver, B.C., 1981), Lecture Notes in Math., 956, Berlin: Springer, pp. 34–71, doi:10.1007/BFb0101508, MR 0704986
- Falb, Peter (1990). Methods of Algebraic Geometry in Control Theory Part II Multivariable Linear Systems and Projective Algebraic Geometry. Springer. ISBN 978-0-8176-4113-9
- Zariski, Oscar (1943), "Foundations of a general theory of birational correspondences.", Trans. Amer. Math. Soc., 53 (3): 490–542, doi:10.2307/1990215, MR 0008468
- Allen Tannenbaum (1982), Invariance and Systems Theory: Algebraic and Geometric Aspects, Lecture Notes in Mathematics, volume 845, Springer-Verlag, ISBN 9783540105657
- Zariski, Oscar (1949), "A simple analytical proof of a fundamental property of birational transformations.", Proc. Natl. Acad. Sci. U.S.A., 35 (1): 62–66, doi:10.1073/pnas.35.1.62, JSTOR 88284, MR 0028056
- Tsfasman, Michael A.; Vlăduț, Serge G.; Nogin, Dmitry (1990). Algebraic Geometric Codes Basic Notions. American Mathematical Soc. ISBN 978-0-8218-7520-9
- Shenghao, Sun (2011). "L-series of Artin stacks over finite fields". Algebra & Number Theory. 6: 47–122. arXiv:1008.3689v2. doi:10.2140/ant.2012.6.47

5

Polynomials

A polynomial is an algebraic expression that consists of variables and coefficients. It involves the operations of addition, subtraction, multiplication and exponents. Monomials and binomials are polynomials with one and two terms respectively. This chapter discusses in detail the different kinds of polynomials to develop a comprehensive understanding.

Polynomial

When you see and algebraic expression having letters mixed with numbers and arithmetic, like $7x^4-3x^3+19x^2-8x+197$, there is a good chance that it is a polynomial. Mathematicians, scientists, and engineers use polynomials to solve problems. Polynomials are taught in algebra, which is a gateway course to all technical subjects.

In algebra, when you see letters, numbers, and arithmetic, the understanding is that the letters stand either for numbers you do not know (and may want to find) or for numbers that change during the course of the problem, variables, such as time. A polynomial is an algebraic expression in which the only arithmetic is addition, subtraction, multiplication, and whole number exponents,. If harder operations are used, such as division or square roots, then this algebraic expression is not a polynomial. Polynomials are often easier to use than other algebraic expressions.

Polynomials are often used to form polynomial equations, such as the equation $7x^4-3x^3+19x^2-8x+197 = 0$, or polynomial functions, such as $f(x) = 7x^4-3x^3+19x^2-8x+197$.

About Polynomials

In a polynomial, "multiplication is understood". That means, for example, that $2x$ means two times x, or twice x. If x is 7, then $2x$ is 14.

The parts of a polynomial separated by plus or minus signs are called "terms". The plus sign or minus sign are part of the term. Thus, in the polynomial x^2+2x-7, the terms are:

$7x^4$

$-3x^3$

$+19x^2$

$-8x$

$+197$

If a polynomial has only one term, it is called a "monomial". $5x^3$ is a monomial. The multiplier out in front is called a "coefficient", the letter is called an "unknown" or a "variable", and the raised number after the x is called an exponent. On a calculator, and on some computers, instead of putting an exponent above and to the right of the x the symbol ^ is used, so that the monomial above could be written 5*x*^3.

A polynomial with exactly three terms is called a "trinomial".

A polynomial with exactly two terms is called a "binomial".

A term with no letters in it is called a "constant term".

A term with one letter but no exponent is called a "first degree term" or a "linear term".

A term with one letter which has exponent 2 is called a "second degree term" or "quadratic term". A "quadratic equation" is an equation in which the largest exponent on any term is 2.

A term with one letter which has exponent 3 is called a "third degree term" or "cubic term." A "cubic equation" is an equation in which the largest exponent on any term is 3.

A term with one letter which has exponent 4 is called a "fourth degree term" or "quartic term." A "quartic equation" is an equation in which the largest exponent on any term is 4.

A term with one letter which has exponent 5 is called a "fifth degree term" or "quintic term." A "quintic equation" is an equation in which the largest exponent on any term is 5.

A term with one letter which has exponent 6 is called a "sixth degree term" or "sextic term." A "sextic equation" is an equation in which the largest exponent on any term is 6.

Trigonometric Polynomial

In the mathematical subfields of numerical analysis and mathematical analysis, a trigonometric polynomial is a finite linear combination of functions $\sin(nx)$ and $\cos(nx)$ with n taking on the values of one or more natural numbers. The coefficients may be taken as real numbers, for real-valued functions. For complex coefficients, there is no difference between such a function and a finite Fourier series.

Trigonometric polynomials are widely used, for example in trigonometric interpolation applied to the interpolation of periodic functions. They are used also in the discrete Fourier transform.

The term *trigonometric polynomial* for the real-valued case can be seen as using the analogy: the functions $\sin(nx)$ and $\cos(nx)$ are similar to the monomial basis for polynomials. In the complex case the trigonometric polynomials are spanned by the positive and negative powers of e^{ix}.

Formal Definition

Any function T of the form

$$T(x) = a_0 + \sum_{n=1}^{N} a_n \cos(nx) + i\sum_{n=1}^{N} b_n \sin(nx) \qquad (x \in \mathbb{R})$$

with $a_n, b_n \in \mathbb{C}$ for $0 \le n \le N$, is called a complex trigonometric polynomial of degree *N*. Using Euler's formula the polynomial can be rewritten as

$$T(x) = \sum_{n=-N}^{N} c_n e^{inx} \qquad (x \in \mathbb{R}).$$

Analogously, letting $a_n, b_n \in \mathbb{R}$, $0 \le n \le N$ and $a_N \ne 0$ or $b_N \ne 0$, then

$$t(x) = a_0 + \sum_{n=1}^{N} a_n \cos(nx) + \sum_{n=1}^{N} b_n \sin(nx) \qquad (x \in \mathbb{R})$$

is called a real trigonometric polynomial of degree *N*.

Properties

A trigonometric polynomial can be considered a periodic function on the real line, with period some multiple of 2π, or as a function on the unit circle.

A basic result is that the trigonometric polynomials are dense in the space of continuous functions on the unit circle, with the uniform norm; this is a special case of the Stone–Weierstrass theorem. More concretely, for every continuous function *f* and every ε > 0, there exists a trigonometric polynomial *T* such that $|f(z) - T(z)| < \varepsilon$ for all *z*. Fejér's theorem states that the arithmetic means of the partial sums of the Fourier series of *f* converge uniformly to *f*, provided *f* is continuous on the circle, thus giving an explicit way to find an approximating trigonometric polynomial *T*.

A trigonometric polynomial of degree *N* has a maximum of 2*N* roots in any open interval $(a, a + 2\pi)$ with a in R, unless it is the zero function.

Matrix Polynomial

In mathematics, a matrix polynomial is a polynomial with square matrices as variables. Given an ordinary, scalar-valued polynomial

$$P(x) = \sum_{i=0}^{n} a_i x^i = a_0 + a_1 x + a_2 x^2 + \cdots + a_n x^n,$$

this polynomial evaluated at a matrix *A* is

$$P(A) = \sum_{i=0}^{n} a_i A^i = a_0 I + a_1 A + a_2 A^2 + \cdots + a_n A^n,$$

where *I* is the identity matrix.

A matrix polynomial equation is an equality between two matrix polynomials, which holds for the specific matrices in question. A matrix polynomial identity is a matrix polynomial equation which holds for all matrices *A* in a specified matrix ring $M_n(R)$.

Characteristic and Minimal Polynomial

The characteristic polynomial of a matrix A is a scalar-valued polynomial, defined by $p_A(t) = \det(tI - A)$. The Cayley–Hamilton theorem states that if this polynomial is viewed as a matrix polynomial and evaluated at the matrix A itself, the result is the zero matrix: $p_A(A) = 0$. The characteristic polynomial is thus a polynomial which annihilates A.

There is a unique monic polynomial of minimal degree which annihilates A; this polynomial is the minimal polynomial. Any polynomial which annihilates A (such as the characteristic polynomial) is a multiple of the minimal polynomial.

It follows that given two polynomials P and Q, we have $P(A) = Q(A)$ if and only if

$$P^{(j)}(\lambda_i) = Q^{(j)}(\lambda_i) \qquad \text{for } j = 0,\dots,n_i - 1 \text{ and } i = 1,\dots,s,$$

where $P^{(j)}$ denotes the jth derivative of P and $\lambda_1,\dots,\lambda_s$ are the eigenvalues of A with corresponding indices $n_1,\dots,n_s$ (the index of an eigenvalue is the size of its largest Jordan block).

Matrix Geometrical Series

Matrix polynomials can be used to sum a matrix geometrical series as one would an ordinary geometric series,

$$\begin{aligned} S &= I + A + A^2 + \cdots + A^n \\ AS &= A + A^2 + A^3 + \cdots + A^{n+1} \\ (I - A)S &= S - AS = I - A^{n+1} \\ S &= (I - A)^{-1}(I - A^{n+1}) \end{aligned}$$

If $I - A$ is nonsingular one can evaluate the expression for the sum S.

Laurent Polynomial

In mathematics, a Laurent polynomial (named after Pierre Alphonse Laurent) in one variable over a field $\mathbb{F}$ is a linear combination of positive and negative powers of the variable with coefficients in $\mathbb{F}$. Laurent polynomials in X form a ring denoted $\mathbb{F}[X, X^{-1}]$. They differ from ordinary polynomials in that they may have terms of negative degree. The construction of Laurent polynomials may be iterated, leading to the ring of Laurent polynomials in several variables. Laurent polynomials are of particular importance in the study of complex variables.

Definition

A Laurent polynomial with coefficients in a field $\mathbb{F}$ is an expression of the form

$$p = \sum_k p_k X^k, \quad p_k \in \mathbb{F}$$

where X is a formal variable, the summation index k is an integer (not necessarily positive) and only finitely many coefficients p_k are non-zero. Two Laurent polynomials are equal if their coefficients are equal. Such expressions can be added, multiplied, and brought back to the same form by reducing similar terms. Formulas for addition and multiplication are exactly the same as for the ordinary polynomials, with the only difference that both positive and negative powers of X can be present:

$$\left(\sum_i a_i X^i\right) + \left(\sum_i b_i X^i\right) = \sum_i (a_i + b_i) X^i$$

and

$$\left(\sum_i a_i X^i\right) \cdot \left(\sum_j b_j X^j\right) = \sum_k \left(\sum_{i,j:i+j=k} a_i b_j\right) X^k.$$

Since only finitely many coefficients a_i and b_j are non-zero, all sums in effect have only finitely many terms, and hence represent Laurent polynomials.

Properties

- A Laurent polynomial over C may be viewed as a Laurent series in which only finitely many coefficients are non-zero.
- The ring of Laurent polynomials $R[X, X^{-1}]$ is an extension of the polynomial ring $R[X]$ obtained by "inverting X". More rigorously, it is the localization of the polynomial ring in the multiplicative set consisting of the non-negative powers of X. Many properties of the Laurent polynomial ring follow from the general properties of localization.
- The ring of Laurent polynomials is a subring of the rational functions.
- The ring of Laurent polynomials over a field is Noetherian (but not Artinian).
- If R is an integral domain, the units of the Laurent polynomial ring $R[X, X^{-1}]$ have the form uX^k, where u is a unit of R and k is an integer. In particular, if K is a field then the units of $K[X, X^{-1}]$ have the form aX^k, where a is a non-zero element of K.
- The Laurent polynomial ring $R[X, X^{-1}]$ is isomorphic to the group ring of the group Z of integers over R. More generally, the Laurent polynomial ring in n variables is isomorphic to the group ring of the free abelian group of rank n. It follows that the Laurent polynomial ring can be endowed with a structure of a commutative, cocommutative Hopf algebra.

Formal Power Series

In mathematics, a formal power series is a generalization of a polynomial, where the number of terms is allowed to be infinite; this implies giving up the possibility of replacing the variable in the polynomial with an arbitrary number. Thus a formal power series differs from a polynomial in that it may have infinitely many terms, and differs from a power series, whose variables can take on numerical values. One way to view a formal power series is as an infinite ordered sequence of numbers. In this case, the powers of the variable are used only to indicate the order of the coefficients, so that the coefficient of x^5 is the fifth term in the sequence. In combinatorics, formal power series provide representations of numerical sequences and of multisets, and for instance allow concise expressions for recursively defined sequences regardless of whether the recursion can be explicitly solved; this is known as the method of generating functions. More generally, formal power series can include series with any finite number of variables, and with coefficients in an arbitrary ring.

Introduction

A formal power series can be loosely thought of as an object that is like a polynomial, but with infinitely many terms. Alternatively, for those familiar with power series (or Taylor series), one may think of a formal power series as a power series in which we ignore questions of convergence by not assuming that the variable X denotes any numerical value (not even an unknown value). For example, consider the series

$$A = 1 - 3X + 5X^2 - 7X^3 + 9X^4 - 11X^5 + \cdots.$$

If we studied this as a power series, its properties would include, for example, that its radius of convergence is 1. However, as a formal power series, we may ignore this completely; all that is relevant is the sequence of coefficients [1, −3, 5, −7, 9, −11, ...]. In other words, a formal power series is an object that just records a sequence of coefficients. It is perfectly acceptable to consider a formal power series with the factorials [1, 1, 2, 6, 24, 120, 720, 5040, ...] as coefficients, even though the corresponding power series diverges for any nonzero value of X.

Arithmetic on formal power series is carried out by simply pretending that the series are polynomials. For example, if

$$B = 2X + 4X^3 + 6X^5 + \cdots,$$

then we add A and B term by term:

$$A + B = 1 - X + 5X^2 - 3X^3 + 9X^4 - 5X^5 + \cdots.$$

We can multiply formal power series, again just by treating them as polynomials:

$$AB = 2X - 6X^2 + 14X^3 - 26X^4 + 44X^5 + \cdots.$$

Notice that each coefficient in the product AB only depends on a *finite* number of coefficients of A and B. For example, the X^5 term is given by

$$44X^5 = (1 \times 6X^5) + (5X^2 \times 4X^3) + (9X^4 \times 2X).$$

For this reason, one may multiply formal power series without worrying about the usual questions of absolute, conditional and uniform convergence which arise in dealing with power series in the setting of analysis.

Once we have defined multiplication for formal power series, we can define multiplicative inverses as follows. The multiplicative inverse of a formal power series A is a formal power series C such that $AC = 1$, provided that such a formal power series exists. It turns out that if A has a multiplicative inverse, it is unique, and we denote it by A^{-1}. Now we can define division of formal power series by defining B/A to be the product BA^{-1}, provided that the inverse of A exists. For example, one can use the definition of multiplication above to verify the familiar formula

$$\frac{1}{1+X} = 1 - X + X^2 - X^3 + X^4 - X^5 + \cdots.$$

An important operation on formal power series is coefficient extraction. In its most basic form, the coefficient extraction operator $[X^n]$ applied to a formal power series A in one variable extracts the coefficient of the nth power of the variable, so that $[X^2]A = 5$ and $[X^5]A = -11$. Other examples include

$$\begin{aligned}
\left[X^3\right](B) &= 4,\\
\left[X^2\right](X + 3X^2Y^3 + 10Y^6) &= 3Y^3,\\
\left[X^2Y^3\right](X + 3X^2Y^3 + 10Y^6) &= 3,\\
\left[X^n\right]\left(\frac{1}{1+X}\right) &= (-1)^n,\\
\left[X^n\right]\left(\frac{X}{(1-X)^2}\right) &= n.
\end{aligned}$$

Similarly, many other operations that are carried out on polynomials can be extended to the formal power series setting, as explained below.

The Ring of Formal Power Series

The set of all formal power series in X with coefficients in a commutative ring R form another ring that is written $R[[X]]$, and called the ring of formal power series in the variable X over R.

Definition of the Formal Power Series Ring

One can characterize $R[[X]]$ abstractly as the completion of the polynomial ring $R[X]$ equipped with a particular metric. This automatically gives $R[[X]]$ the structure of a topological ring (and even of a complete metric space). But the general construction of a completion of a metric space is more involved than what is needed here, and would make formal power series seem more complicated than they are. It is possible to describe $R[[X]]$ more explicitly, and define the ring structure and topological structure separately, as follows.

Ring Structure

As a set, $R[[X]]$ can be constructed as the set $R^{\mathbb{N}}$ of all infinite sequences of elements of R, indexed by the natural numbers (taken to include 0). Designating a sequence whose term at index n is a_n by (a_n), one defines addition of two such sequences by

$$(a_n)_{n\in\mathbb{N}} + (b_n)_{n\in\mathbb{N}} = (a_n + b_n)_{n\in\mathbb{N}}$$

and multiplication by

$$(a_n)_{n\in\mathbb{N}} \times (b_n)_{n\in\mathbb{N}} = \left(\sum_{k=0}^{n} a_k b_{n-k}\right)_{n\in\mathbb{N}}.$$

This type of product is called the Cauchy product of the two sequences of coefficients, and is a sort of discrete convolution. With these operations, $R^{\mathbb{N}}$ becomes a commutative ring with zero element $(0,0,0,\cdots)$ and multiplicative identity $(1,0,0,\cdots)$.

The product is in fact the same one used to define the product of polynomials in one indeterminate, which suggests using a similar notation. One embeds R into $R[[X]]$ by sending any (constant) $a \in R$ to the sequence $(a,0,0,\cdots)$ and designates the sequence $(0,1,0,0,\cdots)$ by X; then using the above definitions every sequence with only finitely many nonzero terms can be expressed in terms of these special elements as

$$(a_0, a_1, a_2, \cdots, a_n, 0, 0, \ldots) = a_0 + a_1 X + \cdots + a_n X^n = \sum_{i=0}^{n} a_i X^i;$$

these are precisely the polynomials in X. Given this, it is quite natural and convenient to designate a general sequence $(a_n)_{n\in\mathbb{N}}$ by the formal expression $\sum_{i\in\mathbb{N}} a_i X^i$, even though the latter *is not* an expression formed by the operations of addition and multiplication defined above (from which only finite sums can be constructed). This notational convention allows reformulation of the above definitions as

$$\left(\sum_{i\in\mathbb{N}} a_i X^i\right) + \left(\sum_{i\in\mathbb{N}} b_i X^i\right) = \sum_{i\in\mathbb{N}} (a_i + b_i) X^i$$

and

$$\left(\sum_{i\in\mathbb{N}} a_i X^i\right) \times \left(\sum_{i\in\mathbb{N}} b_i X^i\right) = \sum_{n\in\mathbb{N}} \left(\sum_{k=0}^{n} a_k b_{n-k}\right) X^n.$$

which is quite convenient, but one must be aware of the distinction between formal summation (a mere convention) and actual addition.

Topological Structure

Having stipulated conventionally that

$$(a_0, a_1, a_2, a_3, \ldots) = \sum_{i=0}^{\infty} a_i X^i,$$

one would like to interpret the right hand side as a well-defined infinite summation. To that end, a notion of convergence in $R^{\mathbb{N}}$ is defined and a topology on $R^{\mathbb{N}}$ is constructed. There are several equivalent ways to define the desired topology.

- We may give $R^{\mathbb{N}}$ the product topology, where each copy of R is given the discrete topology.
- We may give $R^{\mathbb{N}}$ the I-adic topology, where $I = (X)$ is the ideal generated by X, which consists of all sequences whose first term a_0 is zero.

- The desired topology could also be derived from the following metric. The distance between distinct sequences (a_n) and (b_n) in $R^{\mathbb{N}}$, is defined to be

  $$d((a_n),(b_n)) = 2^{-k},$$

  where k is the smallest natural number such that $a_k \neq b_k$; the distance between two equal sequences is of course zero.

Informally, two sequences $\{a_n\}$ and $\{b_n\}$ become closer and closer if and only if more and more of their terms agree exactly. Formally, the sequence of partial sums of some infinite summation converges if for every fixed power of X the coefficient stabilizes: there is a point beyond which all further partial sums have the same coefficient. This is clearly the case for the right hand side of (1), regardless of the values a_n, since inclusion of the term for $i = n$ gives the last (and in fact only) change to the coefficient of X^n. It is also obvious that the limit of the sequence of partial sums is equal to the left hand side.

This topological structure, together with the ring operations described above, form a topological ring. This is called the ring of formal power series over R and is denoted by $R[[X]]$. The topology has the useful property that an infinite summation converges if and only if the sequence of its terms converges to 0, which just means that any fixed power of X occurs in only finitely many terms.

The topological structure allows much more flexible use of infinite summations. For instance the rule for multiplication can be restated simply as

$$\left(\sum_{i\in\mathbb{N}} a_i X^i\right)\times\left(\sum_{i\in\mathbb{N}} b_i X^i\right) = \sum_{i,j\in\mathbb{N}} a_i b_j X^{i+j},$$

since only finitely many terms on the right affect any fixed X^n. Infinite products are also defined by the topological structure; it can be seen that an infinite product converges if and only if the sequence of its factors converges to 1.

Alternative Topologies

The above topology is the finest topology for which $\sum_{i=0}^{\infty} a_i X^i$ always converges as a summation to the formal power series designated by the same expression, and it often suffices to give a meaning to infinite sums and products, or other kinds of limits that one wishes to use to designate particular formal power series. It can however happen occasionally that one wishes to use a coarser topology, so that certain expressions become convergent that would otherwise diverge. This applies in particular when the base ring R already comes with a topology other than the discrete one, for instance if it is also a ring of formal power series.

Consider the ring of formal power series

$$\mathbb{Z}[[X]][[Y]]$$

then the topology of above construction only relates to the indeterminate Y, since the topology that was put on $\mathbb{Z}[[X]]$ has been replaced by the discrete topology when defining the topology of the whole ring. So

$$\sum_{i\in\mathbb{N}} XY^i$$

converges to the power series suggested, which can be written as $\frac{X}{1-Y}$; however the summation

$$\sum_{i\in\mathbb{N}} X^i Y$$

would be considered to be divergent, since every term affects the coefficient of Y (which coefficient is itself a power series in X). This asymmetry disappears if the power series ring in Y is given the product topology where each copy of $\mathbb{Z}[[X]]$ is given its topology as a ring of formal power series rather than the discrete topology. As a consequence, for convergence of a sequence of elements of $\mathbb{Z}[[X]][[Y]]$ it then suffices that the coefficient of each power of Y converges to a formal power series in X, a weaker condition that stabilizing entirely; for instance in the second example given here the coefficient of Y converges to $\frac{1}{1-X}$, so the whole summation converges to $\frac{Y}{1-X}$.

This way of defining the topology is in fact the standard one for repeated constructions of rings of formal power series, and gives the same topology as one would get by taking formal power series in all inderteminates at once. In the above example that would mean constructing $\mathbb{Z}[[X,Y]]$, and here a sequence converges if and only if the coefficient of every monomial X^iY^j stabilizes. This topology, which is also the I-adic topology, where $I=(X,Y)$ is the ideal generated by X and Y, still enjoys the property that a summation converges if and only if its terms tend to 0.

The same principle could be used to make other divergent limits converge. For instance in $\mathbb{R}[[X]]$ the limit

$$\lim_{n\to\infty}\left(1+\frac{X}{n}\right)^n$$

does not exist, so in particular it does not converge to $\exp(X)=\sum_{n\in\mathbb{N}}\frac{X^n}{n!}$. This is because for $i\geq 2$ the coefficient $\binom{n}{i}/n^i$ of X^i does not stabilize as $n\to\infty$. It does however converge in the usual topology of $\mathbb{R}$, and in fact to the coefficient $\frac{1}{i!}$ of $\exp(X)$. Therefore, if one would give $\mathbb{R}[[X]]$ the product topology of $\mathbb{R}^{\mathbb{N}}$ where the topology of $\mathbb{R}$ is the usual topology rather than the discrete one, then the above limit would converge to $\exp(X)$. This more permissive approach is not however the standard when considering formal power series, as it would lead to convergence considerations that are as subtle as they are in analysis, while the philosophy of formal power series is on the contrary to make convergence questions as trivial as they can possibly be. With this topology it would *not* be the case that a summation converges if and only if its terms tend to 0.

Universal Property

The ring $R[[X]]$ may be characterized by the following universal property. If S is a commutative associative algebra over R, if I is an ideal of S such that the I-adic topology on S is complete, and if x is an element of I, then there is a *unique* $\Phi: R[[X]]\to S$ with the following properties:

- Φ is an R-algebra homomorphism

- Φ is continuous
- $\Phi(X) = x$.

Operations on Formal Power Series

One can perform algebraic operations on power series to generate new power series. Besides the ring structure operations defined above, we have the following.

Power Series Raised to Powers

If n is a natural number we have

$$\left(\sum_{k=0}^{\infty} a_k X^k\right)^n = \sum_{m=0}^{\infty} c_m X^m,$$

where

$$c_0 = a_0^n, \quad c_m = \frac{1}{ma_0}\sum_{k=1}^{m}(kn - m + k)a_k c_{m-k},$$

for $m \geq 1$. (This formula can only be used if m and a_0 are invertible in the ring of scalars.)

In the case of formal power series with complex coefficients, the complex powers are well defined at least for series f with constant term equal to 1. In this case, f^α can be defined either by composition with the binomial series $(1+x)^\alpha$, or by composition with the exponential and the logarithmic series, $f^\alpha := \exp(\alpha\log(f))$, or as the solution of the differential equation $f(f^\alpha)' = \alpha f^\alpha f'$ with constant term 1, the three definitions being equivalent. The rules of calculus $(f^\alpha)^\beta = f^{\alpha\beta}$ and $f^\alpha g^\alpha = (fg)^\alpha$ easily follow.

Inverting Series

The series

$$A = \sum_{n=0}^{\infty} a_n X^n$$

in $R[[X]]$ is invertible in $R[[X]]$ if and only if its constant coefficient a_0 is invertible in R. This condition is necessary, for the following reason: if we suppose that A has an inverse $B = b_0 + b_1 x + \cdots$ then the constant term $a_0 b_0$ of $A \cdot B$ is the constant term of the identity series, i.e., it is 1. This condition is also sufficient; we may compute the coefficients of the inverse series B via the explicit recursive formula

$$b_0 = \frac{1}{a_0}$$

$$b_n = -\frac{1}{a_0}\sum_{i=1}^{n} a_i b_{n-i} \qquad \textit{for } n \geq 1.$$

An important special case is that the geometric series formula is valid in $K[[X]]$:

$$(1-X)^{-1}=\sum_{n=0}^{\infty}X^n.$$

If $R=K$ is a field, then a series is invertible if and only if the constant term is non-zero, i.e., if and only if the series is not divisible by X. This says that $K[[X]]$ is a discrete valuation ring with uniformizing parameter X.

Dividing Series

The computation of a quotient $f/g=h$

$$\frac{\sum_{n=0}^{\infty}b_nX^n}{\sum_{n=0}^{\infty}a_nX^n}=\sum_{n=0}^{\infty}c_nX^n,$$

assuming the denominator is invertible (that is, a_0 is invertible in the ring of scalars), can be performed as a product f and the inverse of g, or directly equating the coefficients in $f=gh$:

$$c_n=\frac{1}{a_0}\left(b_n-\sum_{k=1}^{n}a_kc_{n-k}\right).$$

Extracting Coefficients

The coefficient extraction operator applied to a formal power series

$$f(X)=\sum_{n=0}^{\infty}a_nX^n$$

in X is written

$$\left[X^m\right]f(X)$$

and extracts the coefficient of X^m, so that

$$\left[X^m\right]f(X)=\left[X^m\right]\sum_{n=0}^{\infty}a_nX^n=a_m.$$

Composition of series

Given formal power series

$$f(X)=\sum_{n=1}^{\infty}a_nX^n=a_1X+a_2X^2+\cdots$$

$$g(X)=\sum_{n=0}^{\infty}b_nX^n=b_0+b_1X+b_2X^2+\cdots,$$

one may form the *composition*

$$g(f(X)) = \sum_{n=0}^{\infty} b_n (f(X))^n = \sum_{n=0}^{\infty} c_n X^n,$$

where the coefficients c_n are determined by "expanding out" the powers of $f(X)$:

$$c_n := \sum_{k \in \mathbb{N}, |j|=n} b_k a_{j_1} a_{j_2} \cdots a_{j_k}.$$

Here the sum is extended over all (k, j) with k in N and $j \in \mathbb{N}_+^k$ with $|j| := j_1 + \cdots + j_k = n.$

A more explicit description of these coefficients is provided by Faà di Bruno's formula, at least in the case where the coefficient ring is a field of characteristic 0.

A point here is that this operation is only valid when $f(X)$ has *no constant term*, so that each c_n depends on only a finite number of coefficients of $f(X)$ and $g(X)$. In other word the series for $g(f(X))$ converges in the topology of $R[[X]]$.

Example

Assume that the ring R has characteristic 0. If we denote by $\exp(X)$ the formal power series

$$\exp(X) = 1 + X + \frac{X^2}{2!} + \frac{X^3}{3!} + \frac{X^4}{4!} + \cdots,$$

then the expression

$$\exp(\exp(X) - 1) = 1 + X + X^2 + \frac{5X^3}{6} + \frac{5X^4}{8} + \cdots$$

makes perfect sense as a formal power series. However, the statement

$$\exp(\exp(X)) = e\exp(\exp(X) - 1) = e + eX + eX^2 + \frac{5eX^3}{6} + \cdots$$

is not a valid application of the composition operation for formal power series. Rather, it is confusing the notions of convergence in $R[[X]]$ and convergence in R; indeed, the ring R may not even contain any number e with the appropriate properties.

Composition Inverse

Whenever a formal series $f(X) = \sum_k f_k X^k \in R[[X]]$ has $f_0 = 0$ and f_1 being an invertible element of R, there exists a series $g(X) = \sum_k g_k X^k$ that is the composition inverse of f, meaning that composing f with g gives the series representing the identity function (whose first coefficient is 1 and all other coefficients are zero). The coefficients of g may be found recursively by using the above formula for the coefficients of a composition, equating them with those of the composition identity

X (that is 1 at degree 1 and 0 at every degree greater than 1) . In the case when the coefficient ring is a field of characteristic 0, the Lagrange inversion formula provides a powerful tool to compute the coefficients of g, as well as the coefficients of the (multiplicative) powers of g.

Formal Differentiation of Series

Given a formal power series

$$f = \sum_{n\geq 0} a_n X^n$$

in $R[[X]]$, we define its formal derivative, denoted Df or f', by

$$Df = \sum_{n\geq 1} a_n n X^{n-1}.$$

The symbol D is called the formal differentiation operator. The motivation behind this definition is that it simply mimics term-by-term differentiation of a polynomial.

This operation is R-linear:

$$D(af + bg) = a \cdot Df + b \cdot Dg$$

for any a, b in R and any f, g in $R[[X]]$. Additionally, the formal derivative has many of the properties of the usual derivative of calculus. For example, the product rule is valid:

$$D(fg) = f \cdot (Dg) + (Df) \cdot g,$$

and the chain rule works as well:

$$D(f \circ g) = (Df \circ g) \cdot Dg,$$

whenever the appropriate compositions of series are defined.

Thus, in these respects formal power series behave like Taylor series. Indeed, for the f defined above, we find that

$$(D^k f)(0) = k! a_k,$$

where D^k denotes the kth formal derivative (that is, the result of formally differentiating k times).

Properties

Algebraic Properties of the Formal Power Series Ring

$R[[X]]$ is an associative algebra over R which contains the ring $R[X]$ of polynomials over R; the polynomials correspond to the sequences which end in zeros.

The Jacobson radical of $R[[X]]$ is the ideal generated by X and the Jacobson radical of R; this is implied by the element invertibility criterion discussed above.

The maximal ideals of $R[[X]]$ all arise from those in R in the following manner: an ideal M of $R[[X]]$ is maximal if and only if $M \cap R$ is a maximal ideal of R and M is generated as an ideal by X and $M \cap R$.

Several algebraic properties of R are inherited by $R[[X]]$:

- if R is a local ring, then so is $R[[X]]$;
- if R is Noetherian, then so is $R[[X]]$; this is a version of the Hilbert basis theorem;
- if R is an integral domain, then so is $R[[X]]$;
- if K is a field, then $K[[X]]$ is a discrete valuation ring.

Topological Properties of the Formal Power Series Ring

The metric space $(R[[X]], d)$ is complete.

The ring $R[[X]]$ is compact if and only if R is finite. This follows from Tychonoff's theorem and the characterisation of the topology on $R[[X]]$ as a product topology.

Applications

Formal power series can be used to solve recurrences occurring in number theory and combinatorics. For an example involving finding a closed form expression for the Fibonacci numbers, Examples of generating functions.

One can use formal power series to prove several relations familiar from analysis in a purely algebraic setting. Consider for instance the following elements of Q[[X]]:

$$\sin(X) := \sum_{n\geq 0} \frac{(-1)^n}{(2n+1)!} X^{2n+1}$$

$$\cos(X) := \sum_{n\geq 0} \frac{(-1)^n}{(2n)!} X^{2n}$$

Then one can show that

$$\sin^2(X) + \cos^2(X) = 1,$$

$$\frac{\partial}{\partial X} \sin(X) = \cos(X),$$

$$\sin(X+Y) = \sin(X)\cos(Y) + \cos(X)\sin(Y).$$

The last one being valid in the ring Q[[X,Y]].

For K a field, the ring $K[[X_1, ..., X_r]]$ is often used as the "standard, most general" complete local ring over K in algebra.

Interpreting Formal Power Series as Functions

In mathematical analysis, every convergent power series defines a function with values in the real or complex numbers. Formal power series can also be interpreted as functions, but one has to be careful with the domain and codomain. If $f = \sum a_n X^n$ is an element of $R[[X]]$, S is a commutative associative algebra over R, I is an ideal in S such that the I-adic topology on S is complete, and x is an element of I, then we can define

$$f(x) = \sum_{n\geq 0} a_n x^n.$$

This latter series is guaranteed to converge in S given the above assumptions on x. Furthermore, we have

$$(f+g)(x) = f(x) + g(x)$$

and

$$(fg)(x) = f(x)g(x).$$

Unlike in the case of bona fide functions, these formulas are not definitions but have to be proved.

Since the topology on $R[[X]]$ is the (X)-adic topology and $R[[X]]$ is complete, we can in particular apply power series to other power series, provided that the arguments don't have constant coefficients (so that they belong to the ideal (X)): $f(0)$, $f(X^2-X)$ and $f((1-X)^{-1} - 1)$ are all well defined for any formal power series $f \in R[[X]]$.

With this formalism, we can give an explicit formula for the multiplicative inverse of a power series f whose constant coefficient $a = f(0)$ is invertible in R:

$$f^{-1} = \sum_{n\geq 0} a^{-n-1}(a-f)^n.$$

If the formal power series g with $g(0) = 0$ is given implicitly by the equation

$$f(g) = X$$

where f is a known power series with $f(0) = 0$, then the coefficients of g can be explicitly computed using the Lagrange inversion formula.

Generalizations

Formal Laurent Series

A formal Laurent series over a ring R is defined in a similar way to a formal power series, except that we also allow finitely many terms of negative degree (this is different from the classical Laurent series), that is series of the form

$$f = \sum_{n\in\mathbb{Z}} a_n X^n$$

where $a_n = 0$ for all but finitely many negative indices n. Multiplication of such series can be defined. Indeed, similarly to the definition for formal power series, the coefficient of X^k of two series with respective sequences of coefficients $\{a_n\}$ and $\{b_n\}$ is

$$\sum_{i\in\mathbb{Z}} a_i b_{k-i},$$

which sum is effectively finite because of the assumed vanishing of coefficients at sufficiently negative indices, and which sum zero for sufficiently negative k for the same reason.

For a non-zero formal Laurent series, the minimal integer n such that $a_n \neq 0$ is called the order of f, denoted $\operatorname{ord}(f)$. (The order of the zero series is $+\infty$.) The formal Laurent series form the ring of formal Laurent series over R, denoted by $R((X))$. It is equal to the localization of $R[[X]]$ with respect to the set of positive powers of X. It is a topological ring with the metric $d(f,g) = 2^{-ord(f-g)}$.

If $R = K$ is a field, then $K((X))$ is in fact a field, which may alternatively be obtained as the field of fractions of the integral domain $K[[X]]$.

One may define formal differentiation for formal Laurent series in a natural way (term-by-term). Precisely, the formal derivative of the formal Laurent series f above is

$$f' = Df = \sum_{n\in\mathbb{Z}} na_n X^{n-1}$$

which is again an element of $K((X))$. Notice that if f is a non-constant formal Laurent series, and K is a field of characteristic 0, then one has

$$ord(f') = ord(f) - 1.$$

However, in general this is not the case since the factor *n* for the lowest order term could be equal to 0 in *R*.

Formal Residue

Assume that K is a field of characteristic 0. Then the map

$$D : K((X)) \to K((X))$$

is a K-derivation that satisfies

$$\ker D = K$$

$$\operatorname{im} D = \{f \in K((X)) : [X^{-1}]f = 0\}.$$

The latter shows that the coefficient of X^{-1} in f is of particular interest; it is called *formal residue of* f and denoted $Res(f)$. The map

$$\operatorname{Res} : K((X)) \to K$$

is K -linear, and by the above observation one has an exact sequence

$$0 \to K \to K((X)) \xrightarrow{D} K((X)) \xrightarrow{Res} K \to 0.$$

Some rules of calculus. As a quite direct consequence of the above definition, and of the rules of formal derivation, one has, for any $f, g \in K((X))$

i. $\text{Res}(f') = 0;$

ii. $\text{Res}(fg') = -\text{Res}(f'g);$

iii. $\text{Res}(f'/f) = \text{ord}(f), \quad \forall f \neq 0;$

iv. $\text{Res}((f \circ g)g') = \text{ord}(g)\text{Res}(f)$, if $\text{ord}(g) > 0;$

v. $[X^n]f(X) = \text{Res}\left(X^{-n-1}f(X)\right).$

Property (i) is part of the exact sequence above. Property (ii) follows from (i) as applied to $(fg)' = f'g + fg'$. Property (iii): any f can be written in the form $f = x^m g$, with $m = ord(f)$ and $\text{ord}(g) = 0$: then $f'/f = mX^{-1} + g'/g$. Since $\text{ord}(g) = 0$, the element g is invertible in $K[[X]] \subset im(D) = ker(Res)$, whence $\text{Res}(f'/f) = m$. Property (iv): Since $\text{im}(D) = ker(Res)$, we can write $f = f_{-1}X^{-1} + F'$, with $F \in K[[X]]$. Consequently, $(f \circ g)g' = f_{-1}g^{-1}g' + (F' \circ g)g' = f_{-1}g'/g + (F \circ g)'$and (iv) follows from (i) and (iii). Property (v) is clear from the definition.

The Lagrange Inversion Formula

As mentioned above, any formal series $f \in K[[X]]$ with $f_0 = 0$ and $f_1 \neq 0$ has a composition inverse g in $K[[X]]$. The following relation between the coefficients of g^n and f^{-k} holds ("Lagrange inversion formula"):

$$k[X^k]g^n = n[X^{-n}]f^{-k}.$$

In particular, for $n = 1$ and all $k \geq 1$,

$$[X^k]g = \frac{1}{k} Res\left(f^{-k}\right).$$

Since the proof of the Lagrange inversion formula is a very short computation, it is worth reporting it here. Since $ord(f) = 1$, by the above rules of calculus,

$$k\left[X^k\right]g^n = k\,\text{Res}\left(g^n X^{-k-1}\right) = k\text{ord}(f)\text{Res}\left(g^n X^{-k-1}\right) = k\text{Res}\left(\left(\left(g^n X^{-k-1}\right) o f\right) f'\right)$$

$$= k\,\text{Res}\left((g\,o\,f)^n f^{-k-1} f'\right) = k\text{Res}\left(X^n f^{-k-1} f'\right) = -\text{Res}\left(X^n \left(f^{-k}\right)'\right)$$

$$= \text{Res}\left(\left(X^n\right)' f^{-k}\right) = n\,\text{Res}\left(X^{n-1-k}\right) = n\left[X^{-n}\right]f^{-k}.$$

Generalizations. One may observe that the above computation can be repeated plainly in more general settings than $K((X))$: a generalization of the Lagrange inversion formula is already available working in the $C((X))$-modules $X^\alpha C((X))$, where α is a complex exponent. As a consequence, if

f and g are as above, with $f_1 = g_1 = 1$, we can relate the complex powers of f/X and g/X: precisely, if α and β are non-zero complex numbers with negative integer sum, $m=-\alpha-\beta \in \mathbb{N}$, then

$$\frac{1}{\alpha}[X^m]\left(\frac{f}{X}\right)^{\alpha} = -\frac{1}{\beta}[X^m]\left(\frac{g}{X}\right)^{\beta}.$$

For instance, this way one finds the power series for complex powers of the Lambert function.

Power Series in Several Variables

Formal power series in any number of indeterminates (even infinitely many) can be defined. If I is an index set and X_I is the set of indeterminates X_i for $i \in I$, then a monomial X^α is any finite product of elements of X_I (repetitions allowed); a formal power series in X_I with coefficients in a ring R is determined by any mapping from the set of monomials X^α to a corresponding coefficient c_α, and is denoted $\sum_\alpha c_\alpha X^\alpha$. The set of all such formal power series is denoted $R[[X_I]]$, and it is given a ring structure by defining

$$\left(\sum_\alpha c_\alpha X^\alpha\right) + \left(\sum_\alpha d_\alpha X^\alpha\right) = \sum_\alpha (c_\alpha + d_\alpha) X^\alpha$$

and

$$\left(\sum_\alpha c_\alpha X^\alpha\right) \times \left(\sum_\beta d_\beta X^\beta\right) = \sum_{\alpha,\beta} c_\alpha d_\beta X^{\alpha+\beta}$$

Topology

The topology on $R[[X_I]]$ is such that a sequence of its elements converges only if for each monomial X^α the corresponding coefficient stabilizes. If I is finite, then this the J-adic topology, where J is the ideal of $R[[X_I]]$ generated by all the indeterminates in X_I. This does not hold if I is infinite. For example, if $I = \mathbb{N}$, then the sequence $(f_n)_{n \in \mathbb{N}}$ with $f_n = X_n + X_{n+1} + X_{n+2} + \dots$ does not converge with respect to any J-adic topology on R, but clearly for each monomial the corresponding coefficient stabilizes.

As remarked above, the topology on a repeated formal power series ring like $R[[X]][[Y]]$ is usually chosen in such a way that it becomes isomorphic as a topological ring to $R[[X,Y]]$.

Operations

All of the operations defined for series in one variable may be extended to the several variables case.

- A series is invertible if and only if its constant term is invertible in R.
- The composition $f(g(X))$ of two series f and g is defined if f is a series in a single indeterminate, and the constant term of g is zero. For a series f in several indeterminates a form of "composition" can similarly be defined, with as many separate series in the place of g as there are indeterminates.

In the case of the formal derivative, there are now separate partial derivative operators, which differentiate with respect to each of the indeterminates. They all commute with each other.

Universal Property

In the several variables case, the universal property characterizing $R[[X_1, ..., X_r]]$ becomes the following. If S is a commutative associative algebra over R, if I is an ideal of S such that the I-adic topology on S is complete, and if $x_1, ..., x_r$ are elements of I, then there is a *unique* $\Phi : R[[X_1, ..., X_n]] \to S$ with the following properties:

- Φ is an R-algebra homomorphism
- Φ is continuous
- $\Phi(X_i) = x_i$ for $i = 1, ..., r$.

Non-commuting Variables

The several variable case can be further generalised by taking *non-commuting variables* X_i for $i \in I$, where I is an index set and then a monomial X^α is any word in the X_I; a formal power series in X_I with coefficients in a ring R is determined by any mapping from the set of monomials X^α to a corresponding coefficient c_α, and is denoted $\sum_\alpha c_\alpha X^\alpha$. The set of all such formal power series is denoted $R«X_I»$, and it is given a ring structure by defining addition pointwise

$$\left(\sum_\alpha c_\alpha X^\alpha\right)+\left(\sum_\alpha d_\alpha X^\alpha\right)=\sum_\alpha (c_\alpha + d_\alpha)X^\alpha$$

and multiplication by

$$\left(\sum_\alpha c_\alpha X^\alpha\right)\times\left(\sum_\alpha d_\alpha X^\alpha\right)=\sum_{\alpha,\beta} c_\alpha d_\beta X^\alpha \cdot X^\beta$$

where $\cdot$ denotes concatenation of words. These formal power series over R form the Magnus ring over R.

On a Semiring

In theoretical computer science, the following definition of a formal power series is given: let Σ be an alphabet (finite set) and S be a semiring. In this context, a formal power series is any mapping r from the set of strings generated by Σ (denoted as Σ^*) to the semiring S. The values of such a mapping r are (somewhat idiosyncratically) denoted as (r, w) were $w \in \Sigma^*$. Then the mapping r itself is conventionally written as $r = \sum_{w\in\Sigma^*} (r,w)w$. Given this notation, the values (r, w) are also called the coefficients of the series. Similarly to the non-commuting [ring] case discussed in the section above this, the notation for the collection of all power series given a fixed alphabet and semiring is $S\langle\langle\Sigma^*\rangle\rangle$.

Replacing the Index set by an Ordered Abelian Group

Suppose G is an ordered abelian group, meaning an abelian group with a total ordering < respecting the group's addition, so that $a<b$ if and only if $a+c<b+c$ for all c. Let I be a well-ordered

subset of G, meaning I contains no infinite descending chain. Consider the set consisting of

$$\sum_{i \in I} a_i X^i$$

for all such I, with a_i in a commutative ring R, where we assume that for any index set, if all of the a_i are zero then the sum is zero. Then $R((G))$ is the ring of formal power series on G; because of the condition that the indexing set be well-ordered the product is well-defined, and we of course assume that two elements which differ by zero are the same. Sometimes the notation $[[R^G]]$ is used to denote $R((G))$.

Various properties of R transfer to $R((G))$. If R is a field, then so is $R((G))$. If R is an ordered field, we can order $R((G))$ by setting any element to have the same sign as its leading coefficient, defined as the least element of the index set I associated to a non-zero coefficient. Finally if G is a divisible group and R is a real closed field, then $R((G))$ is a real closed field, and if R is algebraically closed, then so is $R((G))$.

This theory is due to Hans Hahn, who also showed that one obtains subfields when the number of (non-zero) terms is bounded by some fixed infinite cardinality.

Examples and Related Topics

- Bell series are used to study the properties of multiplicative arithmetic functions
- Formal groups are used to define an abstract group law using formal power series
- Puiseux series are an extension of formal Laurent series, allowing fractional exponents
- Rational series

Monomial

In mathematics, a monomial is, roughly speaking, a polynomial which has only one term. Two definitions of a monomial may be encountered:

- (1): A monomial, also called power product, is a product of powers of variables with nonnegative integer exponents, or, in other words, a product of variables, possibly with repetitions. The constant 1 is a monomial, being equal to the empty product and x^0 for any variable x. If only a single variable x is considered, this means that a monomial is either 1 or a power x^n of x, with n a positive integer. If several variables are considered, say, x, y, z then each can be given an exponent, so that any monomial is of the form x^a, y^b, z^c with a, b, c non-negative integers (taking note that any exponent 0 makes the corresponding factor equal to 1).
- (2): A monomial is a monomial in the first sense multiplied by a nonzero constant, called the coefficient of the monomial. A monomial in the first sense is a special case of a monomial in the second sense, where the coefficient is 1. For example, in this interpretation $-7x^5$ and $(3-4i)x^4yz^{13}$ are monomials (in the second example, the variables are $x, y, z,$ and the coefficient is an imaginary complex number).

In the context of Laurent polynomials and Laurent series, the exponents of a monomial may be negative, and in the context of Puiseux series, the exponents may be rational numbers.

Since the word "monomial", as well as the word "polynomial", comes from the late Latin word "binomium" (binomial), by changing the prefix "bi" (two in Latin), a monomial should theoretically be called a "mononomial". "Monomial" is a syncope by haplology of "mononomial".

Comparison of the Two Definitions

With either definition, the set of monomials is a subset of all polynomials that is closed under multiplication.

Both uses of this notion can be found, and in many cases the distinction is simply ignored, see for instance examples for the first and second meaning. In informal discussions the distinction is seldom important, and tendency is towards the broader second meaning. When studying the structure of polynomials however, one often definitely needs a notion with the first meaning. This is for instance the case when considering a monomial basis of a polynomial ring, or a monomial ordering of that basis. An argument in favor of the first meaning is also that no obvious other notion is available to designate these values (the term power product is in use, in particular when *monomial* is used with the first meaning, but it does not make the absence of constants clear either), while the notion term of a polynomial unambiguously coincides with the second meaning of monomial.

Monomial Basis

The most obvious fact about monomials (first meaning) is that any polynomial is a linear combination of them, so they form a basis of the vector space of all polynomials, called the *monomial basis* - a fact of constant implicit use in mathematics.

Number

The number of monomials of degree d in n variables is the number of multicombinations of d elements chosen among the n variables (a variable can be chosen more than once, but order does not matter), which is given by the multiset coefficient $\left(\!\!\binom{n}{d}\!\!\right)$. This expression can also be given in the form of a binomial coefficient, as a polynomial expression in d, or using a rising factorial power of $d + 1$:

$$\left(\!\!\binom{n}{d}\!\!\right) = \binom{n+d-1}{d} = \binom{d+(n-1)}{n-1} = \frac{(d+1)\times(d+2)\times\cdots\times(d+n-1)}{1\times 2\times\cdots\times(n-1)} = \frac{1}{(n-1)!}(d+1)^{\overline{n-1}}.$$

The latter forms are particularly useful when one fixes the number of variables and lets the degree vary. From these expressions one sees that for fixed n, the number of monomials of degree d is a polynomial expression in d of degree $n-1$ with leading coefficient $\frac{1}{(n-1)!}$.

For example, the number of monomials in three variables ($n=3$) of degree d is $\frac{1}{2}(d+1)^{\overline{2}} = \frac{1}{2}(d+1)(d+2)$; these numbers form the sequence 1, 3, 6, 10, 15, ... of triangular numbers.

The Hilbert series is a compact way to express the number of monomials of a given degree: the number of monomials of degree d in n variables is the coefficient of degree d of the formal power series expansion of

$$\frac{1}{(1-t)^n}.$$

The number of monomials of degree at most d in n variables is $\binom{n+d}{n} = \binom{n+d}{d}$. This follows from the one-to-one correspondence between the monomials of degree d in $n+1$ variables and the monomials of degree at most d in n variables, which consists in substituting by 1 the extra variable.

Notation

Notation for monomials is constantly required in fields like partial differential equations. If the variables being used form an indexed family like x_1, x_2, x_3, ..., then *multi-index notation* is helpful: if we write

$$\alpha = (a,b,c)$$

we can define

$$x^\alpha = x_1^a \, x_2^b \, x_3^c$$

and save a great deal of space.

Degree

The degree of a monomial is defined as the sum of all the exponents of the variables, including the implicit exponents of 1 for the variables which appear without exponent; e.g., in the example of the previous section, the degree is $a+b+c$. The degree of xyz^2 is 1+1+2=4. The degree of a nonzero constant is 0. For example, the degree of -7 is 0.

The degree of a monomial is sometimes called order, mainly in the context of series. It is also called total degree when it is needed to distinguish it from the degree in one of the variables.

Monomial degree is fundamental to the theory of univariate and multivariate polynomials. Explicitly, it is used to define the degree of a polynomial and the notion of homogeneous polynomial, as well as for graded monomial orderings used in formulating and computing Gröbner bases. Implicitly, it is used in grouping the terms of a Taylor series in several variables.

Geometry

In algebraic geometry the varieties defined by monomial equations $x^\alpha = 0$ for some set of α have special properties of homogeneity. This can be phrased in the language of algebraic groups, in terms of the existence of a group action of an algebraic torus (equivalently by a multiplicative group of diagonal matrices). This area is studied under the name of *torus embeddings*.

Binomial

In algebra, a binomial is a polynomial that is the sum of two terms, each of which is a monomial. It is the simplest kind of polynomial after the monomials.

Definition

A binomial is a polynomial which is the sum of two monomials. A binomial in a single indeterminate (also known as a univariate binomial) can be written in the form

$$ax^m - bx^n,$$

where a and b are numbers, and m and n are distinct nonnegative integers and x is a symbol which is called an indeterminate or, for historical reasons, a variable. In the context of Laurent polynomials, a *Laurent binomial*, often simply called a *binomial*, is similarly defined, but the exponents m and n may be negative.

More generally, a binomial may be written as:

$$ax_1^{n_1} \cdots x_i^{n_i} - bx_1^{m_1} \cdots x_i^{m_i}$$

Some examples of binomials are:

$$3x - 2x^2$$

$$xy + yx^2$$

$$0.9x^3 + \pi y^2$$

Operations on Simple Binomials

- The binomial $x^2 - y^2$ can be factored as the product of two other binomials:

  $$x^2 - y^2 = (x+y)(x-y).$$

  This is a special case of the more general formula:

  $$x^{n+1} - y^{n+1} = (x-y)\sum_{k=0}^{n} x^k y^{n-k}.$$

  When working over the complex numbers, this can also be extended to:

  $$x^2 + y^2 = x^2 - (iy)^2 = (x-iy)(x+iy).$$

- The product of a pair of linear binomials $(ax + b)$ and $(cx + d)$ is a trinomial:

  $$(ax+b)(cx+d) = acx^2 + (ad+bc)x + bd.$$

- A binomial raised to the nth power, represented as $(x + y)^n$ can be expanded by means of the binomial theorem or, equivalently, using Pascal's triangle. For example, the square

$(x + y)^2$ of the binomial $(x + y)$ is equal to the sum of the squares of the two terms and twice the product of the terms, that is:

$$(x+y)^2 = x^2 + 2xy + y^2.$$

The numbers (1, 2, 1) appearing as multipliers for the terms in this expansion are binomial coefficients two rows down from the top of Pascal's triangle. The expansion of the nth power uses the numbers n rows down from the top of the triangle.

- An application of above formula for the square of a binomial is the "(m, n)-formula" for generating Pythagorean triples:

  For $m < n$, let $a = n^2 - m^2$, $b = 2mn$, and $c = n^2 + m^2$; then $a^2 + b^2 = c^2$.

- Binomials that are sums or differences of cubes can be factored into lower-order polynomials as follows:

  $$x^3 + y^3 = (x+y)(x^2 - xy + y^2)$$
  $$x^3 - y^3 = (x-y)(x^2 + xy + y^2)$$

Homogeneous Polynomial

In mathematics, a homogeneous polynomial is a polynomial whose nonzero terms all have the same degree. For example, $x^5 + 2x^3y^2 + 9xy^4$ is a homogeneous polynomial of degree 5, in two variables; the sum of the exponents in each term is always 5. The polynomial $x^3 + 3x^2y + z^7$ is not homogeneous, because the sum of exponents does not match from term to term. A polynomial is homogeneous if and only if it defines a homogeneous function. An algebraic form, or simply form, is a function defined by a homogeneous polynomial. A binary form is a form in two variables. A *form* is also a function defined on a vector space, which may be expressed as a homogeneous function of the coordinates over any basis.

A polynomial of degree 0 is always homogeneous; it is simply an element of the field or ring of the coefficients, usually called a constant or a scalar. A form of degree 1 is a linear form. A form of degree 2 is a quadratic form. In geometry, the Euclidean distance is the square root of a quadratic form.

Homogeneous polynomials are ubiquitous in mathematics and physics. They play a fundamental role in algebraic geometry, as a projective algebraic variety is defined as the set of the common zeros of a set of homogeneous polynomials.

Properties

A homogeneous polynomial defines a homogeneous function. This means that, if a multivariate polynomial P is homogeneous of degree d, then

$$P(\lambda x_1, \ldots, \lambda x_n) = \lambda^d P(x_1, \ldots, x_n),$$

for every λ in any field containing the coefficients of P. Conversely, if the above relation is true for infinitely many λ then the polynomial is homogeneous of degree d.

In particular, if P is homogeneous then

$$P(x_1,\ldots,x_n)=0 \;\Rightarrow\; P(\lambda x_1,\ldots,\lambda x_n)=0,$$

for every λ This property is fundamental in the definition of a projective variety.

Any nonzero polynomial may be decomposed, in a unique way, as a sum of homogeneous polynomials of different degrees, which are called the homogeneous components of the polynomial.

Given a polynomial ring $R=K[x_1,\ldots,x_n]$ over a field (or, more generally, a ring) K, the homogeneous polynomials of degree d form a vector space (or a module), commonly denoted R_d. The above unique decomposition means that R is the direct sum of the R_d (sum over all nonnegative integers).

The dimension of the vector space (or free module) R_d is the number of different monomials of degree d in n variables (that is the maximal number of nonzero terms in a homogeneous polynomial of degree d in n variables). It is equal to the binomial coefficient

$$\binom{d+n-1}{n-1}=\binom{d+n-1}{d}=\frac{(d+n-1)!}{d!(n-1)!}.$$

Homogeneous polynomial satisfy Euler's identity for homogeneous functions. That is, if P is a homogeneous polynomial of degree d in the indeterminates $x_1,\ldots,x_n$, one has, whichever is the commutative ring of the coefficients,

$$dP=\sum_{i=1}^{n} x_i \frac{\partial P}{\partial x_i},$$

where $\frac{\partial P}{\partial x_i}$ denotes the formal partial derivative of P with respect to x_i .

Homogenization

A non-homogeneous polynomial $P(x_1,\ldots,x_n)$ can be homogenized by introducing an additional variable x_0 and defining the homogeneous polynomial sometimes denoted ${}^hP$:

$${}^hP(x_0,x_1,\ldots,x_n)=x_0^d P\left(\frac{x_1}{x_0},\ldots,\frac{x_n}{x_0}\right),$$

where d is the degree of P. For example, if

$$P=x_3^3+x_1x_2+7,$$

then

$${}^hP=x_3^3+x_0x_1x_2+7x_0^3.$$

A homogenized polynomial can be dehomogenized by setting the additional variable $x_0 = 1$. That is

$$P(x_1,\ldots,x_n)={}^hP(1,x_1,\ldots,x_n).$$

Quasi-homogeneous Polynomial

In algebra, a multivariate polynomial

$$f(x) = \sum_{\alpha} a_\alpha x^\alpha, \ where\ \alpha = (i_1, \ldots, i_r) \in \mathbb{N}^r, \text{ and } x^\alpha = x_1^{i_1} \cdots x_r^{i_r},$$

is quasi-homogeneous or weighted homogeneous, if there exist r integers $w_1, \ldots, w_r$, called weights of the variables, such that the sum $w = w_1 i_1 + \cdots + w_r i_r$ is the same for all nonzero terms of *f*. This sum *w* is the *weight* or the *degree* of the polynomial.

The term *quasi-homogeneous* comes from the fact that a polynomial *f* is quasi-homogeneous if and only if

$$f(\lambda^{w_1} x_1, \ldots, \lambda^{w_r} x_r) = \lambda^w f(x_1, \ldots, x_r)$$

for every λ in any field containing the coefficients.

A polynomial $f(x_1, \ldots, x_n)$ is quasi-homogeneous with weights $w_1, \ldots, w_r$ if and only if

$$f(y_1^{w_1}, \ldots, y_n^{w_n})$$

is a homogeneous polynomial in the y_i. In particular, a homogeneous polynomial is always quasi-homogeneous, with all weights equal to 1.

In other words, a polynomial is quasi-homogeneous if all the α belong to the same affine hyperplane. As the Newton polytope of the polynomial is the convex hull of the set $\{\alpha \mid a_\alpha \neq 0\}$, the quasi-homogeneous polynomials may also be defined as the polynomials that have a degenerate Newton polytope (here "degenerate" means "contained in some affine hyperplane").

Introduction

Consider the polynomial $f(x, y) = 5x^3y^3 + xy^9 - 2y^{12}$. This one has no chance of being a homogeneous polynomial; however if instead of considering $f(\lambda x, \lambda y)$ we use the pair (λ^3, λ) to test *homogeneity*, then

$$f(\lambda^3 x, \lambda y) = 5(\lambda^3 x)^3 (\lambda y)^3 + (\lambda^3 x)(\lambda y)^9 - 2(\lambda y)^{12} = \lambda^{12} f(x, y).$$

We say that $f(x, y)$ is a quasi-homogeneous polynomial of type (3,1), because its three pairs (i_1, i_2) of exponents (3,3), (1,9) and (0,12) all satisfy the linear equation $3i_1 + 1i_2 = 12$. In particular, this says that the Newton polytope of $f(x, y)$ lies in the affine space with equation $3x + y = 12$ inside $\mathbb{R}^2$.

The above equation is equivalent to this new one: $\frac{1}{4}x + \frac{1}{12}y = 1$. Some authors prefer to use this last condition and prefer to say that our polynomial is quasi-homogeneous of type $(\frac{1}{4}, \frac{1}{12})$.

As noted above, a homogeneous polynomial $g(x, y)$ of degree *d* is just a quasi-homogeneous polynomial of type (1,1); in this case all its pairs of exponents will satisfy the equation $1i_1 + 1i_2 = d$.

Definition

Let $f(x)$ be a polynomial in r variables $x = x_1 \ldots x_r$ with coefficients in a commutative ring R. We express it as a finite sum

$$f(x) = \sum_{\alpha \in \mathbb{N}^r} a_\alpha x^\alpha, \alpha = (i_1, \ldots, i_r), a_\alpha \in \mathbb{R}.$$

We say that f is quasi-homogeneous of type $\varphi = (\varphi_1, \ldots, \varphi_r)$, $\varphi_i \in \mathbb{N}$ if there exists some $a \in \mathbb{R}$ such that

$$\langle \alpha, \varphi \rangle = \sum_k^r i_k \varphi_k = a,$$

whenever $a_\alpha \neq 0$.

Multi-homogeneous Bézout Theorem

In algebra and algebraic geometry, the multi-homogeneous Bézout theorem is a generalization to multi-homogeneous polynomials of Bézout's theorem, which counts the number of isolated common zeros of a set of homogeneous polynomials. This generalization is due to Igor Shafarevich.

A multi-homogeneous polynomial is a polynomial that is homogeneous with respect to several sets of variables.

More precisely, let us consider k positive integers $n_1, \ldots, n_k$, and, for i = 1, ..., k, the $n_i + 1$ indeterminates $x_{i,0}, x_{i,1}, \ldots, x_{i,n}$. A polynomial in all these indeterminates is multi-homogeneous of multi-degree $d_1, \ldots, d_k$, if it is homogeneous of degree d_i in $x_{i,0}, x_{i,1}, \ldots, x_{i,n}$.

A multi-projective variety is a projective subvariety of the product of projective spaces

$$\mathbb{P}_{n_1} \times \cdots \times \mathbb{P}_{n_k},$$

where $\mathbb{P}_n$ denote the projective space of dimension n. A multi-projective variety may be defined as the set of the common nontrivial zeros of an ideal of multi-homogeneous polynomials, where "nontrivial" means that $x_{i,0}, x_{i,1}, \ldots, x_{i,n}$ are not simultaneously 0, for each i.

Bézout's theorem asserts that n homogeneous polynomials of degree $d_1, \ldots, d_n$ in $n + 1$ indeterminates define either an algebraic set of positive dimension, or a zero-dimensional algebraic set consisting of $d_1 \cdots d_n$ points counted with their multiplicities.

For stating the generalization of Bézout's theorem, it is convenient to introduce new indeterminates $t_1, \ldots, t_k$, and to represent the multi-degree $d_1, \ldots, d_k$ by the linear form $\mathbf{d} = d_1 t_1 + \cdots + d_k t_k$. In the following, "multi-degree" will refer to this linear form rather than to the sequence of degrees.

Setting $n = n_1 + \cdots + n_k$, the multi-homogeneous Bézout theorem is the following.

With above notation, n multi-homogeneous polynomials of multi-degrees $\mathbf{d}_1, \ldots, \mathbf{d}_n$ *define either an multi-projective algebraic set of positive dimension, or a zero-dimensional algebraic set consisting of B points, counted with multiplicities, where B is the coefficient of*

$$t_1^{n_1} \cdots t_k^{n_k}$$

in the product of linear forms

$$\mathrm{d}_1 \cdots \mathrm{d}_n.$$

Gauss's Lemma

In algebra, in the theory of polynomials (a subfield of ring theory), Gauss's lemma is either of two related statements about polynomials with integer coefficients:

- The first result states that the product of two primitive polynomials is primitive (a polynomial with integer coefficients is called *primitive* if the greatest common divisor of its coefficients is 1).
- The second result states that if a non-constant polynomial with integer coefficients is irreducible over the integers, then it is also irreducible if it is considered as a polynomial over the rationals.

This second statement is a consequence of the first. The first statement and proof of the lemma are in Article 42 of Carl Friedrich Gauss's *Disquisitiones Arithmeticae* (1801). These statements have several generalizations described below.

Primitive and Irreducible Polynomials

The notion of primitive polynomial used here (which differs from the notion with the same name in the context of finite fields) is defined in any polynomial ring $R[X]$ where R is a commutative ring: a polynomial P in $R[X]$ is primitive if the only elements of R that divide all coefficients of P at once are the invertible elements of R. In the case where R is the ring Z of the integers, this is equivalent to the condition that no prime number divides all the coefficients of P. The notion of irreducible element is defined in any integral domain: an element is irreducible if it is not invertible and cannot be written as a product of two non-invertible elements. If R is an integral domain then so is the polynomial ring $R[X]$ because the leading coefficient of product of non-zero polynomials in $R[X]$ is equal to the product of their leading coefficients, hence is nonzero. A non-constant irreducible polynomial in $R[X]$ is one that is not a product of two non-constant polynomials *and* which is primitive (because being primitive excludes precisely non-invertible constant polynomials as factors). Note that an irreducible element of R is still irreducible when viewed as constant polynomial in $R[X]$; this explains the need for "non-constant" above, and in the irreducibility statements below.

Version Valid Over Integers

The two properties of polynomials with integer coefficients can now be formulated formally as follows:

Primitivity statement: The set of primitive polynomials in Z[*X*] is closed under multiplication: if *P* and *Q* are primitive polynomials then so is their product *PQ*.

Proof: Suppose the product of two primitive polynomials $f(x)$ and $g(x)$ is not primitive, so there exists a prime number p that is a common divisor of all the coefficients of the product. But since $f(x)$ and $g(x)$ are primitive, p cannot divide either all the coefficients of $f(x)$ or all those of $g(x)$. Let $a_r x^r$ and $b_s x^s$ be the first (i.e., highest degree) terms with a coefficient not divisible by p, respectively in $f(x)$ and in $g(x)$. Now consider the coefficient of x^{r+s} in the product. Its value is given by $\sum a_i b_j$, where the sum runs over all pairs of indices i,j such that $i + j = r + s$. This sum contains a term $a_r b_s$ which is not divisible by p (because p is prime, by Euclid's lemma), yet all the remaining ones are (because either $i < r$ or $j < s$), so the entire sum is not divisible by p. But by assumption *all* coefficients in the product are divisible by p, leading to a contradiction. Therefore, the coefficients of the product can have no common divisor and are thus primitive. This completes the proof.

Alternatively the statement can be proved as a special case of more general results given below.

Irreducibility statement: A non-constant polynomial in Z[*X*] is irreducible in Z[*X*] if and only if it is both irreducible in Q[*X*] and primitive in Z[*X*].

This is a special case for R = Z of more general irreducibility statement proved below.

The second result implies that if a polynomial with integer coefficients can be factored over the rational numbers, then there exists a factorization over the integers. This fact is often useful when combined with results such as Eisenstein's criterion.

An application is the rational root theorem.

The second result also implies that the minimal polynomial over the rational numbers of an algebraic integer has integer coefficients.

Version Valid Over any GCD Domain

Gauss's lemma holds more generally over arbitrary GCD domains. There the *content* $c(P)$ of a polynomial P can be defined as the greatest common divisor of the coefficients of P (like the gcd, the content is actually a class of associate elements).

Primitivity statement: If R is a GCD domain, then the set of primitive polynomials in $R[X]$ is closed under multiplication. More generally, the content of a product ST of polynomials is the product $c(S)c(T)$ of their contents.

Proof: The latter part follows from the former since $c(S)c(T)$ is certainly a common divisor of the coefficients of the product, so one can divide by $c(S)$ and $c(T)$ to reduce S and T to primitive polynomials. For the proof of the former part we proceed by induction on the total number of nonzero terms of S and T combined. If one of the polynomials has at most one term, the result is obvious;

this covers in particular all cases with fewer than 4 nonzero terms. So let both S and T have at least 2 terms, and assume the result established for any smaller combined number of terms. By dividing S by $c(S)$ and T by $c(T)$, we reduce to the case $c(S) = c(T) = 1$. If the content $c = c(ST)$ is not invertible, it has a non-trivial divisor in common with the leading coefficient of at least one of S and T (since it divides their product, which is the leading coefficient of ST). Suppose by symmetry that this is the case for S, let L be the leading term of S, and let $d = \gcd(c,c(L))$ be the mentioned common divisor (here the content $c(L)$ of L is just its unique coefficient). Since d is a common divisor of ST and LT, it also divides $(S - L)T$, in other words it divides its content, which by induction (since $S - L$ has fewer terms than S) is $c(S - L)c(T) = c(S - L)$. As d also divides $c(L)$, it divides $c(S) = 1$, which gives a contradiction; therefore $c(ST)$ is invertible (and can be taken to be 1).

Irreducibility statement: Let R be a GCD domain and F its field of fractions. A non-constant polynomial in $R[X]$ is irreducible in $R[X]$ if and only if it is both irreducible in $F[X]$ and primitive in $R[X]$.

Proof: As mentioned above a non-constant polynomial is irreducible in $R[X]$ if and only if it is primitive and not a product of two non-constant polynomials in $R[X]$. Being irreducible in $F[X]$ certainly excludes the latter possibility (since those non-constant polynomials would remain non-invertible in $F[X]$), so the essential point left to prove is that if P is non-constant and irreducible in $R[X]$ then it is irreducible in $F[X]$. Note first that in $F[X]\setminus\{0\}$ any class of associate elements (whose elements are related by multiplication by nonzero elements of the field F) meets the set of primitive elements in $R[X]$: starting from an arbitrary element of the class, one can first (if necessary) multiply by a nonzero element of R to enter into the subset $R[X]$ (removing denominators), then divide by the greatest common divisor of all coefficients to obtain a primitive polynomial. Now assume that P is reducible in $F[X]$, so $P = ST$ with S,T non-constant polynomials in $F[X]$. One can replace S and T by associate primitive elements S', T', and obtain $P = \alpha S'T'$ for some nonzero α in F. But $S'T'$ is primitive in $R[X]$ by the primitivity statement, so α must lie in R (if α is written as a fraction a/b, then b has to divide all coefficients of $aS'T'$, so b divides $c(aS'T') = a$, which means $\alpha = a/b$ is in R) and the decomposition $P = \alpha S'T'$ contradicts the irreducibility of P in $R[X]$.

The condition that R is a GCD domain is not superfluous because it implies that every irreducible element of this ring is also a prime element, which in turn implies that every nonzero element of R has at most one factorization into a product of irreducible elements and a unit up to order and associate relationship. In a ring where factorization is not unique, say $pa = qb$ with p and q irreducible elements that do not divide any of the factors on the other side, the product $(p + qX)(a + qX) = pa + (p+a)qX + q^2X^2 = q(b + (p+a)X + qX^2)$ shows the failure of the primitivity statement. For a concrete example one can take $R = \mathbb{Z}[i\sqrt{5}]$, $p = 1 + i\sqrt{5}$, $a = 1 - i\sqrt{5}$, $q = 2$, $b = 3$. In this example the polynomial $3 + 2X + 2X^2$ (obtained by dividing the right hand side by $q = 2$) provides an example of the failure of the irreducibility statement (it is irreducible over R, but reducible over its field of fractions $\mathbb{Q}[i\sqrt{5}]$). Another well known example is the polynomial $X^2 - X - 1$, whose roots are the golden ratio $\varphi = (1 + \sqrt{5})/2$ and its conjugate $(1 - \sqrt{5})/2$ showing that it is reducible over the field $\mathbb{Q}[\sqrt{5}]$, although it is irreducible over the non-UFD $\mathbb{Z}[\sqrt{5}]$ which has $\mathbb{Q}[\sqrt{5}]$ as field of fractions. In the latter example the ring can be made into an UFD by taking its integral closure $\mathbb{Z}[\varphi]$ in $\mathbb{Q}[\sqrt{5}]$ (the ring of Dirichlet integers), over which $X^2 - X - 1$ becomes reducible, but in the former example R is already integrally closed.

In the special case when R is unique factorization domain (UFD), the primitivity statement can be proved more easily:

Proof: Let S,T be primitive polynomials in $R[X]$, and assume that their product ST is not primitive, so that some non-invertible element d of R divides all coefficients of ST. There is some irreducible element p of R that divides d (this is where we use that R is UFD, if R was only GCD domain then such element wouldn't necessarily exist), and it is also a prime element in R (since R is GCD domain). Then the principal ideal pR generated by p is a prime ideal, so R/pR is an integral domain, and $(R/pR)[X]$ is therefore an integral domain as well (because nonzero polynomials over an integral domain cannot be zero divisors by consideration of the leading coefficient of their product). By hypothesis the projection $R[X] \to (R/pR)[X]$ sends ST to 0, and since this is the product of the projections of S and T (projection is a ring homomorphism), at least one of those projections is 0 (here one uses that $(R/pR)[X]$ is an integral domain). But this means that p divides all of the coefficients either of S or of T, which contradicts its assumed primitivity.

This special case is also sufficient in many applications. For example both primitivity and irreducibility statement for UFD are essential in proving that if R is UFD, then so is $R[X]$ (and by an immediate induction, so is the polynomial ring over R in any number of indeterminates). Any factorization of a polynomial P in $R[X]$ can be split into its irreducible factors that are contained in R (the "constant" factors) and those that are not. The primitivity statement implies that the product Q of the latter (non-constant) irreducible factors is primitive, so the product of the former (constant) factors give the content $c(P)$ of P. This reduces proving uniqueness of factorizations to proving it individually for $c(P)$ and for Q. Since the factorization of $c(P)$ takes place in R, it is unique by assumption. By the irreducibility statement, the irreducible factors that occur in any factorization of Q in $R[X]$ are primitive representatives of irreducible factors in a factorization of Q in $F[X]$. But the latter is unique since $F[X]$ is a principal ideal domain and therefore a unique factorization domain. Together this shows that the factorization of P in $R[X]$ is unique.

Version Valid Over Arbitrary Commutative Ring

As explained above, neither primitivity nor irreducibility statement of Gauss's lemma is valid over general integral domains. However there is a variation of the first statement that is valid even for polynomials over any commutative ring R, which replaces primitivity by the stronger property of co-maximality. Call a polynomial P in $R[X]$ *co-maximal* if the ideal of R generated by the coefficients of the polynomial is the full ring R. Clearly every co-maximal polynomial in $R[X]$ is primitive. If R is a Bézout domain (so in particular if it's a principal ideal domain) then also every primitive polynomial in R is co-maximal. However co-maximality is often much more restrictive condition than primitivity even when R is UFD. For example let k be a field and $R = k[Y,Z]$, which is UFD as explained above. Then the polynomial $P = Y + ZX$ is primitive but not co-maximal in $R[X]$ because the ideal (Y,Z) in R generated by the coefficients of P is proper (primitivity simply means that this ideal isn't contained in any proper principal ideal). We have the following variation of Gauss's lemma:

Co-maximality statement: Let R be a commutative ring. Then the product of two co-maximal polynomials in $R[X]$ is co-maximal.

Proof: Let S,T be co-maximal polynomials in $R[X]$, and assume that their product ST is not co-maximal. Then its coefficients generate a proper ideal I, which by Krull's theorem (which depends on the axiom of choice) is contained in a maximal ideal m of R. Then R/m is a field, and $(R/m)[X]$ is therefore an integral domain. By hypothesis the projection $R[X] \to (R/m)[X]$ sends ST to 0, thus also at least one of S,T individually, which means that its coefficients all lie in m, which contradicts the fact that they generate the whole ring as an ideal.

Permissions

All chapters in this book are published with permission under the Creative Commons Attribution Share Alike License or equivalent. Every chapter published in this book has been scrutinized by our experts. Their significance has been extensively debated. The topics covered herein carry significant information for a comprehensive understanding. They may even be implemented as practical applications or may be referred to as a beginning point for further studies.

We would like to thank the editorial team for lending their expertise to make the book truly unique. They have played a crucial role in the development of this book. Without their invaluable contributions this book wouldn't have been possible. They have made vital efforts to compile up to date information on the varied aspects of this subject to make this book a valuable addition to the collection of many professionals and students.

This book was conceptualized with the vision of imparting up-to-date and integrated information in this field. To ensure the same, a matchless editorial board was set up. Every individual on the board went through rigorous rounds of assessment to prove their worth. After which they invested a large part of their time researching and compiling the most relevant data for our readers.

The editorial board has been involved in producing this book since its inception. They have spent rigorous hours researching and exploring the diverse topics which have resulted in the successful publishing of this book. They have passed on their knowledge of decades through this book. To expedite this challenging task, the publisher supported the team at every step. A small team of assistant editors was also appointed to further simplify the editing procedure and attain best results for the readers.

Apart from the editorial board, the designing team has also invested a significant amount of their time in understanding the subject and creating the most relevant covers. They scrutinized every image to scout for the most suitable representation of the subject and create an appropriate cover for the book.

The publishing team has been an ardent support to the editorial, designing and production team. Their endless efforts to recruit the best for this project, has resulted in the accomplishment of this book. They are a veteran in the field of academics and their pool of knowledge is as vast as their experience in printing. Their expertise and guidance has proved useful at every step. Their uncompromising quality standards have made this book an exceptional effort. Their encouragement from time to time has been an inspiration for everyone.

The publisher and the editorial board hope that this book will prove to be a valuable piece of knowledge for students, practitioners and scholars across the globe.

Index

U

V